T3-BWG-761

LONGITUDINAL STUDIES OF CHILD PERSONALITY

LONGITUDINAL STUDIES OF CHILD PERSONALITY

Abstracts with Index

Alan A. Stone, M.D. and
Gloria Cochrane Onqué, M.D.

Published for THE COMMONWEALTH FUND
By HARVARD UNIVERSITY PRESS
Cambridge, Massachusetts, 1959

Published for
The Commonwealth Fund
By Harvard University Press
Cambridge, Massachusetts

For approximately a quarter of a century THE COMMONWEALTH FUND, through its Division of Publications, sponsored, edited, produced, and distributed books and pamphlets germane to its purposes and operations as a philanthropic foundation. On July 1, 1951, the Fund entered into an arrangement by which HARVARD UNIVERSITY PRESS became the publisher of Commonwealth Fund books, assuming responsibility for their production and distribution. The Fund continues to sponsor and edit its books, and cooperates with the Press in all phases of manufacture and distribution.

Distributed in Great Britain
By Oxford University Press
London

LIBRARY OF CONGRESS CATALOG CARD NO. 59-7646

MANUFACTURED IN THE UNITED STATES OF AMERICA

To Ernst Kris whose enthusiasm launched this project
and
To Milton J. E. Senn whose patience sustained it

FOREWORD

Research workers in the field of personality development and behavior sooner or later feel that their investigations fall short of making the contributions they had hoped because they were too short in range or episodic and because they followed the plans and used the methods of a single scientific discipline. The temptation is great to follow up a study of organisms which are growing and changing, not only to satisfy one's curiosity about their development but also to check on the validity of predictions which such studies invariably imply or actually make. Equally tempting is the desire to join forces with other investigators with different training and experience, especially in other disciplines, in order to make as complete and comprehensive an appraisal as possible.

These were the motives of our group which was drawn from a variety of fields for a coordinated and integrated study of personality development of a group of human beings experiencing the common but critical events of pregnancy and child rearing.

Of necessity this was to be a long-term investigation marked by repeated observations on the *same* individuals — thus a longitudinal study as contrasted to a cross-sectional appraisal in which either a single observation is made on individuals who have similar or different characteristics, or repeated observations are made at various points in time on different groups of individuals. Any study of the phenomena of human development is benefited if it is longitudinal since each observation and measurement of a growing organism has special relevance to that organism at particular times and places. For example, the relationship between illness or a behavioral trait and a specific antecedent event can be accurately determined only by successive observations on the same individual. Furthermore, a longitudinal study, by focusing on an individual throughout his life or on significant segments thereof, permits appraisal of the processes involved in his interactions and transactions with other persons as well as of the resultant behavior and personality. The ideal longitudinal study is a series of observations so spaced as to discover as many of the variabilities as possible occurring during critical periods in the life of an individual so that ultimately predictions of change are possible and correlations between measurements at successive ages are valid. Of course, all things being equal, the meaningfulness and generality of such a study is in proportion to the number of individuals so observed.

Longitudinal studies may be short-term or long-term. Also they may proceed forward in what is called prospective longitudinal research or backward in retrospection, as when a body of data already collected is examined. Often studies are made without the investigator being conscious of the longitudinal character of his work, as when a clinician automatically observes a patient for symptoms of an illness or immunologic change after exposure to an infectious agent. The frequency with which the longitudinal approach is deliberately chosen these days bears witness to the importance of this method in research.

When the late Ernst Kris and I began planning our research project, we were unaware of many of the difficulties and dangers of a study which lasts several years. We knew only some of the hazards, such as the potential loss of a large number of the individuals being studied and the difficulties of long-term financing. A cursory review of the literature revealed a few studies already undertaken which were helpful in overall planning and in suggesting ways of avoiding certain pitfalls and duplication of the work of others, but a thorough study of the literature on longitudinal research of personality development seemed desirable and advisable. It was our good fortune to learn about a small group of Yale medical undergraduate students who were meeting regularly to discuss the literature on psychoanalytic psychology. Dr. Kris, with his customary flare for attracting and enthusing students, suggested that they make an extensive review of bibliographical references of longitudinal studies and abstract the pertinent articles with the ultimate aim of publishing an annotated bibliography. Drs. Alan Stone and Gloria Cochrane (Onqué) obtained the approval of the faculty to develop this project as a joint thesis. Despite the many obstacles inherent in simultaneously doing creditable research and studying medicine full time, Drs. Stone and Cochrane completed their thesis in time to meet the requirements for graduation from the Yale School of Medicine and also to be of great assistance to the team of researchers led by Dr. Kris. It seemed to all of us at the Yale Child Study Center that their annotated bibliography would be valuable to other investigators but that to receive wide dissemination it should be published.

Fortunately, the Commonwealth Fund which supported our longitudinal research with money and counsel agreed to publish it with Harvard University Press. Although several publications of the results of our longitudinal study are planned, we are proud that the first of these is by Drs. Stone and Cochrane because it is the work of two medical undergraduates whom we considered as much members of our team as the more experienced researchers. Thus it appropriately marks the beginning of a continuum of long-term efforts by a group of persons devoted to the study of human beings over a span of time.

Milton J. E. Senn, M.D.
Director, Yale University
Child Study Center

INTRODUCTION

THE MEANING OF THE LONGITUDINAL APPROACH

For the purposes of this survey, the longitudinal approach is a "forward-looking approach which aims to study the individual as he passes from one stage to a later, and to gauge the influence of succeeding experiences and circumstances. . . ."[1] More specifically, it consists of repeated measurements of one or more aspects of behavior in the same child over a given period; these measurements, if reliable, will detect elements of change, development, consistency, and inconsistency.

There seems to be no question about the importance and potential value of longitudinal research. In spite of technical difficulties, all camps of psychological opinion agree on the basic need for such research.

Discussing longitudinal studies, Hartmann and Kris write, "If the longitudinal observation in our civilization were to be systemized and the study of life histories were to be combined with that of the crucial situations in Freud's sense, many hunches might be formulated as propositions, and others might be discarded."[2] This sifting of propositions by testing and retesting is a long process, but in reference to one of the more important aspects of longitudinal research, Bach asserts, "There is no substitute for the slow but reliable experimental and longitudinal 'life studies'

[1] W. E. Blatz and H. Bott, *Parents and the Pre-School Child.* Morrow, New York, 1929, p. 252.

[2] H. Hartmann and E. Kris, The Genetic Approach in Psychoanalysis, *Psychoanalytic Study of the Child* (vol. I). International Universities Press, 1945, p. 28.

of the important dynamics of parent-child relationships."[3] Sears restates the defense in other words: ". . . there is a crying need for the results of longitudinal research on personality development. . . . Until such problems as the organization and interrelationships of pregenital impulses have been examined by reference to records obtained *year after year from the same child,* there are going to be serious lacunae in our knowledge of the motivational sources of adult sexuality and dependencies."[4]

Thus there is no argument about the need for such studies. Unfortunately, however, there are technical difficulties inherent in this type of research, and although they do not detract from the value of the method, they have slowed down and made less effective the production of longitudinal studies. These problems can be classed in three main groups: (1) small samples, (2) poor testing devices, and (3) inability to convert theoretic hypotheses into experimental or operational terms which will define the significant variables.

1. Since the interests of longitudinal studies are so comprehensive, of necessity only a few subjects can be studied. As the study progresses it is often converted into a mere collection of case histories which are valuable in themselves, but invalid as statistical or scientific generalizations.

2. The usual personality tests which are better validated and reliable (Rorschach, T.A.T., etc.) are of limited value for the purposes of longitudinal studies. First, there are many important areas of personality development in which their usefulness is marginal; and second, most of these tests are inapplicable to children under 3 years of age who have not developed adequate verbal responses. Because of this inadequacy many researchers have interrupted longitudinal research in order to develop new tests. Others have given up trying to use or develop tests and have turned to trained observers for the accumulation of their data.

3. When tracing the development of a particular aspect of behavior, often the experimenters have been unable to delimit

[3] G. R. Bach, What Can the Child Psychologist Learn from Childhood Memories of Adults?, *American Psychologist,* 6:307 (1951).

[4] R. Sears, *Survey of Objective Studies of Psychoanalytic Concepts,* New York Social Science Research Council, New York, 1943, p. 141.

from the total data those specific events which are causally related to a particular and final state. Consequently, many studies have accumulated incredible amounts of data which defy any degree of organized analysis and have no relation to a specific, experimentally posed hypothesis. Much of this research, therefore, never reaches the manuscript stage; and if it does, it is formulated as impressions of the author or as case histories.

SOURCES OF DATA

Nevertheless much valuable material has been published in this field, and the aim of the present bibliography is to provide a perspective of the overall contribution made by longitudinal research to psychological knowledge. We have attempted to include the studies to which the general definition of the longitudinal approach is applicable. This annotated bibliography is further limited to those longitudinal studies primarily concerned with psychological (emotional and social) behavior in infants and children. Other areas of behavior (notably physiological, motor, and intellectual) are included only when there is a specified explicit or implicit relation to social and/or emotional factors.

Although in a sense case histories are longitudinal studies, we have had to limit this survey to those which consider a group of children, illustrate a specific syndrome or syndrome complex, or present one case as a sample of a larger organized longitudinal research program. Also certain studies usually considered longitudinal have been excluded, mainly parent diaries of children. First, the enormous number and subjective quality of these diaries and, second, the relative difficulty of summarizing such documents have kept us from including them in this book. Darwin, Hall, Piaget, Allport, and many others have published such studies, and Dennis and Dennis have published a summary of forty such diaries to which the reader is referred.[5]

The collection of data began with a review of well-known journals in the field of child develpoment. Abstracts were made of the longitudinal studies in all volumes of *The Journal of Ge-*

[5] W. Dennis and M. Dennis, Infant Development during the First Year as Shown by 40 Biographies, *Psychological Record*, 1:349–361 (1937).

netic Psychology, The American Journal of Orthopsychiatry, Child Development, Child Development Monographs, Genetic Psychology Monographs, Monographs of the Society for Research in Child Development, and *Enfance.* From the bibliographies of these papers names of other articles and books pertinent to this study were obtained. Research centers and individual authors were contacted and many of these provided us with organized and pertinent bibliographies.

After this initial survey of the longitudinal literature, a more systematic exploration was begun utilizing the *Psychological Abstracts* and the *Child Development Abstracts.* We investigated every article described in these abstracts which seemed to be relevant. If not already included in our initial list, these articles were read and summarized and their bibliographies were examined; the latter provided an additional source for cross-checking our material. Close to one thousand books and articles had been studied as a result of this selection. Those articles which fell into the borderline areas created by the broad definition of longitudinal studies and those which duplicated previous works by the same authors were deleted, although in certain cases the decision for or against exclusion was very difficult. We have taken advantage of the editorial comments of the Commonwealth Fund staff and are indebted to the Harvard University Press readers for their suggestions. However, in any bibliography there are apt to be important omissions and this responsibility must belong to the authors.

This bibliography, then, includes all of the pertinent studies in the *Child Development Abstracts,* the *Psychological Abstracts,* and other papers and books which were not in these reference books. The bibliography is complete up to 1955.

These summaries are intended as essential findings of various longitudinal studies and not as a final reference. It is hoped that they will both save time for the interested reader and better direct his later investigations.

A. A. S.
G. C. O.

TABLE OF CONTENTS

ABSTRACTS

1

Ackerly, S. Rebellion and Its Relation to Delinquency and Neurosis in Sixty Adolescents, *American Journal of Orthopsychiatry,* 3:147–160 (1933)

Setting: Not stated

Subjects: 30 delinquents referred by the juvenile court; 30 nondelinquent siblings for parallel study; average age, 14 years

Time span: 1½ years

Meth. of obs. and test.: Psychiatric case histories; medical histories

The 30 delinquents were compared with their 30 siblings. Also, findings were presented by subgroups based on a comparison of subjects with selected characteristics.

COMPLETE GROUP

Findings

At the time of final comparison no differences in intelligence, health, or body build were found between delinquents and nondelinquents. The delinquent child was the more extraverted in 21 of the 30 pairs of subjects. Histories of hypersensitive behavior in infancy were noted for 15 of the delinquents but none of the siblings. There were 12 delinquent children but only 3 siblings who had poor health histories. Emotional imbalance was noted in 14 of the delinquent children but in only 3 siblings, and 5 delinquents as compared with 2 siblings had had psychotic episodes.

Nearly all the delinquents had poorer school records than their siblings.

Author's interpretations

There might have been certain fundamental constitutional differences between the delinquent and nondelinquent children, e.g., hyperactive behavior in psychotic episodes.

SUBGROUP A

A group of 10 delinquents who were rebelling against physical, mental, and environmental handicaps was compared with a group of 10 nonrebellious, nonhandicapped children who were successful in their school work.

Findings

Although least delinquent in the group, 2 of the most heavily handicapped children seemed to give up their struggle; one by attempting suicide and the other by a pronounced infantile regression. In the other 8 rebellious children, effective rebellion and delinquency seemed to coincide with the maintenance of a comparatively healthy ego and good integration.

SUBGROUP B

A group of 10 rebellious children whose chief complaints were directed against the father was compared with a group of 10 nonrebellious siblings.

Findings

All the fathers of the children in this group were noticeably inadequate. The nonrebellious siblings were often aided by the father and accepted him without overt conflict, although some admitted a secret hostility. Although seriously delinquent, the rebels seemed physically healthier, stronger, more active, and more extraverted than their siblings. The superego of these rebellious children seemed to be held in check by increased activity but its influence could be seen in the courting and acceptance of external punishment.

SUBGROUP C

A group of 10 rebellious delinquents whose chief complaints were directed against the mother was compared with a group of 10 nonrebellious siblings.

Findings

In 6 cases the father was dead and in the other 4 he was a very uninfluential figure. All the mothers were self-employed, aggressive, and domineering and tended to favor the sibling. Of the rebellious children, 8 were markedly neurotic and 6 had experienced psychotic episodes. In 4 of these 6 children it seemed that the delinquent behavior was like a healthy protest, leading to a favorable turning point with escape from social isolation and withdrawal. In the 2 schizophrenic children among the siblings there was no recovery.

2

ALDRICH, C. A. The Prevention of Poor Appetite in Children, *Mental Hygiene,* 10:701–711 (1926)

Setting: Pediatric practice

Subjects: 215 mothers with children under 18 months of age

Time span: 5 years

Meth. of obs. and test.: Medical histories; questionnaires

This group of mothers was educated regarding the possible causes of poor appetite in children (infection, emotional reactions, etc.) and was given the following ten instructions: report to physician food refusal which might be caused by infections; treat food refusal by reducing amount given; facilitate weaning by giving occasional bottle from birth; make gradual food changes; avoid overfeeding; let child's appetite guide; let child choose food; let child alone at mealtime; avoid battles; and try to appreciate the fact that there are emotional reasons for resistance. The eating behavior of the children of this group of mothers was compared with the eating behavior of a group of mothers not similarly oriented.

Author's interpretations

The indoctrinated group encountered fewer feeding problems than a control group, and the difficulties they did encounter could be attributed to failure to follow the advice given.

3

ALDRICH, C. A., SUNG, C., and KNOP, C. The Crying of Newly Born Babies; I — The Community Phase, *Journal of Pediatrics*, 26:313–326 (1945)

Setting: Nursery

Subjects: Number not stated

Time span: 30 days (continuous observations for 24 hours per day)

Meth. of obs. and test.: Direct observations made in the nursery; crying episodes noted as to subject, duration, and (if possible) reasons for crying

Findings

Amount and frequency of crying — For each single hour the average baby cried 1.3 to 11.2 minutes; for a whole day the average was 113.2 minutes per baby. The peak of crying was midnight to 2 a.m., the next highest was from 4 to 7 p.m., and the lowest from 10 to 11 a.m. when babies were fed and the nursery was fully staffed. The diurnal variations in the amounts of crying bore reciprocal relationship to the amount of nursing care. Of 720 equidistant instants, no baby cried in 32%; never more than 53% cried simultaneously. So there is less than a .14% chance that 50% of a group of babies will cry simultaneously at a single instant.

Effect of community living — A study of community crying at a single instant shows that crying is an active, individual reaction to some endogenous or exogenous stimulus, not a contagious phenomenon. Community living as such does not seem to increase crying.

Authors' interpretations

The authors believe that pediatric and obstetrical routines in the nursery (on obstetrics floor) should be synchronized so as to take into account the community needs of the babies. More nurses and shorter intervals between feedings are advisable. The distribution of nurses must also be planned with regard to babies' needs. A nurse whose sole duty is that of trouble-shooter is suggested.

4

Aldrich, C. A., Sung, C., and Knop, C. The Crying of Newly Born Babies; II — The Individual Phase, *Journal of Pediatrics,* 27:89–96 (1945)

Setting: Hospital in Rochester, Minnesota

Subjects: 50 infants; present in nursery for 8 days

Time span: 1 month of continuous observation

Meth. of obs. and test.: Continuous observation by 4 trained observers; recorded amount of crying in minutes, estimated causes

Findings

Amount of crying — The time for the baby with least crying was 386 minutes in 8 days, or 48.2 minutes per day; for the baby with most crying it was 1,947 minutes in 8 days, or 243 minutes per day. The average crying per infant was 936 minutes in 8 days, or 117 minutes per day.

Graphic distribution — The total crying time of 50 subjects was seen as a bell-shaped curve. The graph of the crying of an average baby per day for 8 days resulted in a smooth curve; that of any individual infant rose and fell irregularly.

Causes — Attempts were made to relate crying to weight gain and feeding. Good feeders cried 17.8 minutes less per day. Obvious causes for crying were hunger, vomiting, and soiled or wet diapers. Also, the more nursing care they received, the less they cried. The authors emphasized the importance of *unknown reasons* for crying. These were tabulated in minutes and in number of spells. Unknown causes approached hunger as a cause in total minutes of crying, and unknown causes exceeded hunger in number of crying spells.

5

Aldrich, C. A., Sung, C., and Knop, C. The Crying of Newly Born Babies; III — The Early Period at Home, *Journal of Pediatrics,* 27:428–435 (1945)

Setting: Home

Subjects: 42 infants
Time span: 21 days following dismissal from hospital (average)
Meth. of obs. and test.: Questionnaires

Findings

Amount of crying — On an average day, the average baby at home had 4.0 prolonged crying spells; the average newborn infant in a hospital nursery had 11.9 spells.

Causes — Hunger was the most frequent cause. Most babies cried during their daily bath. It was thought that hot weather might have accounted for a small percentage of summer-month crying. Noise and light were the least likely causes for crying at home. Only 14% of all crying was attributed to unknown causes.

Authors' interpretations

The authors feel that babies differ basically in respect to their tendency to cry, but in view of their finding of fewer negative reactions in the home they suggest that further individualization of nursing care might improve the hospital situation.

6

ALLEN, R. M. A Longitudinal Study of Six Rorschach Protocols of a Three-Year-Old Child, *Child Development,* 22:61–69 (1951)
Setting: Not stated
Subject: Author's son
Time span: 1 year
Meth. of obs. and test.: Rorschach Test

Because of the author's intimate knowledge of the child's environment, the interpretations could be related to the child's recent experiences.

Findings

The article demonstrates that this child used a "magic key" response; that is, having thought up one idea, he applied it to all succeeding cards. During the testing the "contrary phase" developed around age 2½; it is traced as it was reflected in the protocols.

7

Ames, L. B. Early Individual Differences in Visual and Motor Behavior Patterns; a Comparative Study of Two Normal Infants by the Method of Cinemanalysis, *Pedagogical Seminary and Journal of Genetic Psychology,* 65:219–226 (1944)

Setting: Child development clinic

Subjects: 2 boys

Time span: 12 years (subjects were first seen at the age of 2 weeks, then monthly between the ages of 8 and 56 weeks, and 5 and 12 years)

Meth. of obs. and test.: Cinemanalysis

The data were analyzed as to "regard only," "regard and contact," and "contact only."

Findings

Differences in regard and contact — The subjects were observed to "regard only" for the first 24 weeks; at 28 weeks "regard and contact" increased and individual differences became apparent. Perhaps related to these differences was the observation that child D consistently moved more quickly than B and also moved toward objects with more persistence at an earlier age.

Child D — This subject appeared to be quick, active, happy, friendly, well adjusted, forceful, alert, and inquisitive. He was emotionally labile and versatile and perceptive of emotions in others. Child D had superior motor skills and was "somatotonic." He was right-handed and right-footed. Child D consistently scored high in "regard and contact."

Child B — This subject was less vivid and less expressive than Child D. He was sturdy, deliberate, and moderately sociable. Child B had motor ineptness and was "viscerotonic." He was right-handed and left-footed. He consistently scored high in "regard only."

8

Ames, L. B. The Sense of Self of Nursery School Children as Manifested by Their Verbal Behavior, *Pedagogical Seminary and Journal of Genetic Psychology,* 81:193–232 (1952)

Setting: Child development clinic
Subjects: Number not stated; age 18 months to 4 years
Time span: Not stated
Meth. of obs. and test.: Author's observations through one-way mirror

Findings

Age 18 Months

Child by himself — For the most part he is self-contained.

Child-teacher interaction — The child uses single word sentences, e.g., "Hello," "Bye-bye," object-naming, single-word requests; points and asks; and asks for help. He responds to a suggestion by obeying or disobeying, or by mimicking. Nonverbal responses are ignoring, grabbing objects from the teacher, helping or following her, and repetitive play.

Child-child interaction — There is almost no verbalization with other children; occasionally, a child will say "No" to indicate a wish for an object that another child has. Since there are no approaches, there are no responsive acts. Physical disregard is as common as physical interaction. One child may push another out of the way almost impersonally. He may grab at an object without specifically noting who is holding it and the other child may let go or tighten his grip almost reflexively. In most groups each child continues to be an individual — a mobile unit with little contact. One child may stare at another, or two children may play near the same object. Group behavior frequently consists of following the teacher, each child being motivated by his own desire for contact with an adult and the group is formed by accident.

Age 21 Months

Child by himself — The child produces a kind of jargon during his solitary play; he sings and makes two-word descriptions of his own activities.

Child-teacher interaction — There are single-word sentences, consisting mainly of single-word requests but also including two-word requests. The child asks for help, asks for things by pointing and calling by name ("dat") and gives single-word information (e.g., "waining"). He is quite responsive to the teacher and mimics her frequently. He may obey, disobey, or assist the teacher. At times he answers simple questions and is acceptant of physical approach.

Child-child interaction — The child shouts "No" to others and there is some interchange of objects and some imitation. Demands on other children are reinforced with single-word statements. For the main part the other children are treated as objects. There is almost no response to activity. Play is parallel but not interactive and there is a kind of inspectional regard of other children. He begins to spend some time watching the activities of others; but he prefers to watch adults rather than other children. Imitation sometimes becomes a social game.

Age 2 Years

Child by himself — He makes sound effects during his individual play and offers audible directions to himself. He is much less self-sufficient, seeks the teacher and tends to play near her.

Child-teacher interaction — Short sentences are used frequently. There are shared responses and requests are made in longer sentences — e.g., the child will tell the teacher to watch him while he does something. Instead of just looking to see if the teacher notices, the child makes the teacher aware of his activity. Frequently he replies to the teacher by imitating her rather than by responding to her. He is quite acceptant of most suggestions unless they involve physical restraint. He tends to seek out the teacher's company and prefers to imitate the teacher rather than other children.

Child-child interaction — There is a great deal of verbalization over property, both to obtain and to maintain it. Although children occasionally sing together, there is very little responsive verbalization. Several may play in a very small area without actually playing together — that is, in a sense they compose physical but not social groups. Objects are often grabbed from other children but actual fights don't begin to develop until age 2½. Aggression per se is minimal. There is very little involvement in a dominance hierarchy. By this age there is some fairly significant imitation of the behavior of other children. Two kinds of groups are formed: the first around an object of common interest (e.g., the teacher), and the second by physical proximity rather than social interaction.

Age 2½ Years

Child by himself — The child is apt to comment on his own acts and to elaborate verbally on his imaginary play, but for the most part he no longer simply talks to himself. There is much less solitary play, and boasting becomes an important part of the child's behavior.

Child-teacher interaction — The very slightest activity is accompanied by verbalization to the teacher, who is by far the most preferred object for verbalization. The child starts the conversation with the teacher and is not amenable to suggestions from her.

Child-child interaction — Frequent statements of ownership are made and fights over ownership begin to develop. There are attempts to domineer with aggressive threats. Tentative requests and the beginning of early cooperative trends are now noticed. In addition there is some general conversation, but most play is individual except when interference or altercations develop. Aggression at this age is more apt to be spontaneous than retaliative. Some children may attempt to gain the attention of others, but groups of children remain uniformly noncooperative. For the most part the members are brought together by common interest in an object or by accidental proximity.

Age 3 Years

Child by himself — The child comments on his play and engages in a sociable form of play.

Child-teacher interaction — The child describes at length his activities to the teacher and there is much divergent conversational matter. He requests help, information, and attention. Also, the child describes his own independent ability and takes on imaginary roles — e.g., mother and father. Usually he initiates talk with the teacher but may respond, especially to imaginative conversation. He looks to the teacher for help and for discipline.

Child-child interaction — Frequently the child argues about property and resorts to name-calling during the dispute. There are many shared activities and aggression is replaced by excluding individuals from the activity. Children often engage in imaginative play with others and sometimes in cooperative play. They tend to engage in silly laughter when playing together and finally begin fitting names to people. There is some general conversation and a great deal of showing off before other children. The child responds by giving or withholding favors and by imitating others. The group structures are quite fluid. The cliques that develop among 3-year-olds, although not well defined, show that the children have begun to adapt to one another's demands. Some cliques are cooperative, some are not. Twosomes tend to last longer than larger cliques. Group play is often based on imitation.

Age 3½ Years

Child by himself — There is very little self-verbalization. The child tends to be very social and most of his individual behavior has its origin in social interchange. He may repeat previous commands or requests to himself. Imaginary companions are occasionally present.

Child-teacher interaction — The child experiences less negativism and independence. He does not stress his differences as often as formerly. He frequently comments on his own activity. Information is interchanged as he explains his own actions in regard to other children to the teacher. At this age he answers the teacher's questions and imitates her choice of words, but now for the first time he talks more with other children than with the teacher. Also, there is less imaginative play with the teacher and more with other children.

Child-child interaction — Children are less insistent on property rights and more willing to share, to make substitutions, and to take turns. They reason and request more frequently than formerly and give many strong, excluding commands. They often engage in imaginative and cooperative play. Favorites are developed and there is less boasting and more positive interaction. For the first time quarrels may be solved at the verbal level. Most of the children play in groups of at least three which may center about a teacher or other object of common interest or may depend on the imagination.

Age 4 Years

Child by himself — Almost all verbalization is social. There are general comments which give evidence of group awareness. There is more constructive solitary play than at age 3½ but social play predominates.

Child-teacher interaction — The child continues to give the teacher information but it is now much more varied. The teacher is excluded from much of the daily activity and the child develops quite individual and charming techniques of interacting with the teacher and other adults.

Child-child interaction — In general, there is much responsive activity and verbalization. The child asks permission to use objects. He talks less to establish his ownership and understands much more without words. He expresses his desire for interaction by saying, "Let's" with friendly commands. The child rebuffs others and calls them by false or make-believe names even though he knows their real names. There are verbal expressions of friendship; but jealousy is also very strong and is expressed verbally. A child may boast to others in silly language. Play is characterized by more pragmatism and less imagination. At this age, groups formed by three or more children tend to be integrated and to last quite a long time, with much interchange and interaction.

Summary of the Development of Sense of Self

Age 1 to 12 months is a period of self-discovery. At 18 months of age the child is predominantly egocentric; most of his social reactions are to adults. By 21 months, although continuing to be egocentric, he experiences a deeper response to adults. At 2 years he relates to others through objects. At 2½ years his sense of self has been established and he begins to relate to others in terms of this sense. By 3 years of age he has developed many relations with other children, and from then on these child-child relations continue to expand in complexity.

9

Ames, L. B., and Ilg, F. L. Developmental Trends in Writing Behavior, *Pedagogical Seminary and Journal of Genetic Psychology,* 79:29–46 (1951)

Setting: Child development clinic

Subjects: School children

Time span: 50 children tested biannually, age 3 to 6 years; 36 children, age 7 years; 26 children, age 8 years; and 16 children, age 9 years

Meth. of obs. and test.: Spontaneous and directed samples of writing; cinema records of writing posture

The visual motor perception of letters was measured in relation to a clear-cut gradient of ages.

Findings

Age 3 years — At this time circles are drawn in a counterclockwise motion, beginning at the top, and there is some imitative writing. A few children can describe some capital letters.

Age 3½ years — Circles are drawn clockwise, beginning at the bottom. The child can print a few capital letters which have a special significance for him.

Age 4 years — Circles are drawn clockwise, beginning at the top, and circular letters may be described either in a horizontal or a vertical position. Letters may be markedly inaccurate but are usually not reversed. They are printed on the page at random. Some of the children can print their names in very large, irregular letters with interrupted lettering.

Age 5 years — Circles are drawn clockwise, beginning at the top. Half the children can print letters. There is a tendency toward reversals and other inaccuracies. Letters may be drawn in two or three parts. Frequently the top of the letter is the guide rather than the bottom. Some of the children can also write numbers. When a child prints his name, the letters are usually small at the beginning and increase in size toward the end.

Age 5½ years — Circles are drawn clockwise, beginning at the top. There are many reversals of letters and the child begins to use phonetic spelling.

Age 6 years — Circles are drawn counterclockwise, beginning

at the top, and from this age on continue to be drawn in this manner. There are still some reversals of letters but by now the writing skill is becoming stabilized.

10

Ames, L. B., Learned, J., Métraux, R. W., and Walker, R. N. *Child Rorschach Responses*. Hoeber, New York, 1952

Setting: Varied

Subjects: 25 boys, 25 girls; 600 other children tested

Time span: 8 years (biannually between the ages of 2 and 6, and annually between the ages of 6 and 10)

Meth. of obs. and test.: Rorschach Test

No complete inquiry was attempted. At the end of the test period each child was asked to choose those he liked best and those he liked least. The "Loosli" scoring technique is used. The findings are offered in four categories.

Findings

Commonly Scored Determinants

At all ages but age 9 whole responses constitute 50% or more of the total response. Large details constitute 42% at age 2, 48% at age 9, and 40% at age 10. Rare details constitute 4% at age 2, 15% at age 6, and 8% at age 10. Human movement and animal movement increase with age, but at every age animal movement exceeds human movement. Inanimate movement occurs variably but is highest at age 7. ΣC increases until age 7 and then decreases. Pure C responses predominate at age 2 and thereafter CF predominates with a relative increase in FC as age increases. Shading responses are never very important but reach their peak at age 7. Clob responses follow a similar trend. A% decreases slightly though not regularly from 55% at age 2 to 49% at age 10. On the other hand, H% increases from 3% at age 2 to 16% at age 10. F% decreases steadily from 90% at age 2 to 52% at age 7, then finally increases to 63% at age 10. F+% increases steadily with age. Refusals decrease with age. Denials and color naming are minimal, the latter reaching a peak at 2½ years of age. P% rises from 10% at age 2 to 25% at age 10.

Percentage of Subjects Using Each of These Variables

Human and animal movement responses are given by more children as age increases. The same is true for inanimate movement with a peak being reached at age 7. It is also notable that color responses tend to increase with age. FC increases till age 9 and then falls off slightly. CF increases until age 6 and then falls off. C increases to age 7 and then falls off. Shading increases to age 7 and then falls off. Clob responses are conspicuous around age 5 but have fallen off by age 9.

Sex Differences

General findings — There are only a few differences related to sex. On the whole boys give more responses and have more W's than girls. Girls have more Dd than boys. Boys give a greater number of animal movement and inanimate movement responses. For the overall age ranges, M for girls is .62 and for boys .80. FM for girls is .90 and for boys 1.17. For girls m is .20 and for boys .39. Girls give more FC than boys. Flower responses are noted more frequently in girls. ΣC increases in boys. Girls respond to form more than boys. F+% is higher in girls. The ages at which sex differences are marked are interesting.

Age 3½ years — There are some marked differences. Boys give more responses than girls in almost all variables. Thus boys' responses include more animal movement, more initial shock, and more positive identification. ΣC and C are higher in boys. Boys give more shading and more variety of content. They make more mention of sex and elimination and the corresponding anatomical parts.

Age 5 years — Boys show more M and higher ΣC. All boys at this age give at least one movement or color response. Of the 25 girls tested in the longitudinal sample at age 5 years, 6 had an experience balance of OM:OΣC. Girls give more responses concerned with contamination but have better form. Their records are neater and more conforming.

Age 7 years — The greatest differences in responses between the sexes are found at this age. The girls' responses are neat, concise, and orderly; they tend to give global responses whereas the boys pick out details, even tiny ones. The boys' records are longer, more involved and complicated, and often have confabulated elaborations. They have a greater amount of M, of color responses, of shading responses, of Clob responses, and of responses involving blood and fire; and they show an increasing preoccupation with decay, damage, and mutilation.

Age 8 years — Girls begins to show evidence of denial, qualification, making initial explanations to the card, and aggression toward the examiner and the card. They suggest many changes. The boys show more movement and more shading; the girls show more color.

Character of Total Rorschach

Age 2 years — The blot is responded to as a whole. "Magic repetition" is frequently used as a solution of the presentation. The average number of responses is 9.6.

Age 2½ years — The blot continues to be responded to as a whole with a wider number of responses. CF is higher than C; the average number of responses is 10.8. The responses are egocentric, stubborn, and rigid, and indicate that the children are unable to adapt to the demands of the outside world.

Age 3 years — There is nearly a full protocol which shows much better form. The average number of responses is 12.9. The amount of movement in the records increases. Also a tendency for weighing and evaluating begins to develop. There is evidence of conformity, empathy, and some self-control.

Age 3½ years — The protocol is now fully discriminative. The largest proportion of global responses (60%) is given at this age. The responses are imaginative, mostly unstable. The child seems to be insecure and vulnerable and has great difficulty in correlating visual percepts and recognizing customary spatial relationships.

Age 4 years — The protocol is extratensive, egocentric, and unmodulated. Responses are exact, imaginative, violent, furious, and resistant. At this age the responses tend to be predictable and stereotyped.

Age 4½ years — The records show a great deal of confusion

and instability. There is a strong but uncontrolled emotionality. The average number of responses is 14.2. The responses are markedly unpredictable and variable, and are frequently contaminated and confabulated. There is a strong tendency toward S and WS.

Age 5 years — The responses remain primarily global. There is a high degree of generalizing ability. There tends to be only one response per card and this leads to near contamination in the scoring. The results show more introversive protocol, color responses are decreasing, and there is a striving for accurate form representation. Responses are focal, factual, close to home, and close to the subject's own body. Generally, the personality seems to be in good equilibrium, calm, smooth, and untroubled.

Age 5½ years — The protocol shows a sensitive, vulnerable, insecure, unpredictable personality in considerable disequilibrium. The child is outgoing, excitable, and impetuous. The responses tend to be rigid and unmodulated, and give evidence of poor interpersonal relations. This foreshadows a kind of behavior which is most pronounced at age 7.

Age 6 years — The protocol reveals a child who is egocentric, stubborn, expansive, violent, and subject to contradictory tendencies. The records show the child to be quite vulnerable though realistic. However, the responses at this age are predictable. Perceptions are becoming increasingly accurate and there is a full weighing of color and movement.

Age 7 years — There are large numbers of Clob and confabulatory responses, and much shading. Responses involving decay, damage, and mutilation are prominent, and M, FM, and m are higher than at any previous time. The protocol characterizes an "inwardized," withdrawn, brooding, morbid, sensitive, critical, perseverative, self-centered, and thoughtful child who, though thus hemmed in, is still able to generalize intelligently.

Age 8 years —The largest proportion of qualified responses is found at this age. Perseveration and confabulation have begun to decline, and there is a very strong drive toward accuracy. F+ is 87%. The blots seem to represent real things. Color responses are reduced. The examiner arouses some aggression, but the protocol is much less troubled. The child seems more expansive, more dramatic, and more interested in people than at age 7.

Age 9 years — Concern with accuracy continues. Details begin to exceed whole responses but there is an intense drive to produce combining wholes. Form determines 67% of the responses, the largest proportion since age 5. Some of the responses would clearly be indicative of neurotic tendencies if they appeared in adult records. These are expressed as evasive map responses, self-conscious anatomic responses, and much shading, Clob, Dd, and Do. CF responses dominate other types of color responses. There in some aggression toward the examiner and the card. Productivity is high. At this age children vary markedly in their responses.

Age 10 years — For the first time, the experience balance is introversive, i.e., 1.70M:1.51ΣC. There are fewer signs of disturbance than at age 9. There is less shading, fewer Clob responses, less inanimate movement, and very few pure color responses. F+% is highest. There are many refusals but on the whole the children are direct, compliant, and uncomplicated.

Authors' interpretations

The Rorschach Test is an effective instrument for assessing the child's personality. There are normative characteristic patterns of behavior which change with age. A child reveals his basic individual structure quite early and retains this structure during the course of repeated Rorschachs. However, the child's developmental status is often more prominent and more notable than his individual personality structure.

11

ANDRUS, R., and HOROWITZ, E. L. The Effects of Nursery School Training; Insecurity Feelings, *Child Development*, 9:169–174 (1938)

Setting: W. P. A. nursery school

Subjects: Number not stated

Time span: Variable

Meth. of obs. and test.: Rating scales

Findings

The longer children had been in nursery school, the more secure they tended to be rated.

12

Arnstein, F. The Growth of Poetic Criteria in Children, *Progressive Education,* 15:25–34 (1938)

Setting: School

Subjects: 61 children in poetry classes; age 7 to 14 years

Time span: 6 years

Meth. of obs. and test.: Dated records and poems by the children

For the purposes of this study the author defines poetry as "any short genuine expression that is the child's record of an experience real or imagined." Many of the children described in the paper began to write poetry before they were 10 years of age.

Findings

Age 10 years — Self-criticism begins to appear. In the process of maturing the child repudiates his more infantile concepts and expressions and he rejects his own earlier poems as he becomes able to evaluate them at a more mature level.

Age 12 years — Interestingly enough, originality as a criterion for critical judgment is late to emerge, first appearing about grades 7 or 8. It is of interest that at about the same time plagiarism becomes a significant factor in the poetry of some children.

Age 13 to 14 years — Around this age, as the appreciation of original observations develops, the child recognizes the value of a personal, individual approach.

Author's interpretations

At first preoccupation with rhyme and metric reckoning tends to force the child into derivative subject matter and distortion of meaning. Eventually, some of the children learn to appreciate rhyme and meter and develop a sense of how to handle them. The author comments that imagery offers a source of pleasure and a method of exploration for children.

13

ARRINGTON, R. E. *Interrelations in the Behavior of Young Children* (Child Development Monographs, no. 8). Teachers College, Columbia University, New York, 1932

Setting: Nursery school

Subjects: 16 children with longitudinal data; age 2 to 3 years

Time span: 1 year

Meth. of obs. and test.: Teacher's observations

The 16 children in the study were observed and ranked by the time-sampling method, with a year intervening to permit comparisons. The ranking was done for several types of activity — e.g., use of material, physical activity, talking, physical contact, laughing, and crying — and self-centered activity which had no apparent relation to the environment. The individual trends of each child were analyzed in terms of the environmental manipulation (uses of material) and social interaction.

Findings

Group consistency — Talking and laughing were the only activities observed in which the majority of the group showed consistency.

Some illustrative examples of individual trends — In 1930 Ivan was rated aggressive and socially inactive. By the following year he showed a decrease in the initiation of physical contact and an increase in the amount of crying and receiving physical contact. Marjorie maintained consistency in all rated categories, except talking to others and physical contact initiated by others in each of which she showed an increase. Vincent displayed a decrease in the use of material and an increase in social and emotional activity. These changes were a reversal of the course followed by the majority of the group. Horace displayed the most consistent behavior pattern in the group with increased talking to self and a general trend indicative of a movement toward non-social behavior. Neal, the youngest member of the group, remained low in his scores for the use of material and high in his ratings for emotional activity, but over the course of the semester showed increases in social activity. Grace showed the most interesting equivalence of activity toward persons and things.

Eunice was considered to be the "material non-social" type; the changes in her behavior showed intensification of the nonsocial pattern. Rebecca exemplified the nonmaterial, social type; her behavior continued in the direction of predominantly social interaction. Albert was considered to be a "non-material, non-social" type; radical changes in his behavior showed an increased use of material although he remained socially undirected.

Author's interpretations

The children in this study showed a wide variety of behavior differences at age 2. When followed over the period of one year, the group tendency was in the direction of evenly divided material and social interest, but individual variations from the trend were quite marked and important to note.

14

ARRINGTON, R. E. *Time-Sampling Studies of Child Behavior* (Psychological Monographs, vol. 51, no. 2). American Psychological Association, Evanston, Ill., 1939

Setting: Nursery school through third grade

Subjects: 120 children

Time span: Variable

Meth. of obs. and test.: 5-minute observations of use of material, kind of material, frequency of use of material, number of distractions, and frequency and range of verbal, physical and social contact, and language

The author used time-sampling data to rate a group of young children in specific categories. By means of this technique she hoped to develop significant indices of the normal behavior of groups and individuals which could be legitimately compared with one another, irrespective of the subject or the observer's situation and which would be more reliable than selective judgment based on random observation. The purpose was to describe behavior rather than to attempt any explanation of it.

Findings

Language — The collection of language data revealed definite developmental patterns. As the children increased in age, they talked more with other children and less with adults. Nonsocial speech decreased markedly with age. Verbal contacts were higher than physical contacts at all ages.

Physical contact — The data revealed no relation between age and the frequency of physical contact. Other factors were presumably responsible for variations in this behavior.

Use of materials — The data revealed that kindergarten children spent more than three-fourths of their time at some sort of work but only half of this time using materials in what could be called a functional way. Distractions in this kind of work averaged 6 times per 5-minute sample. The boys in the group spent more time at work and the girls seemed more easily distracted.

Individual differences — Individual patterns of social and work activity were defined by the author in terms of average frequency of behavior per 5-minute observation and the amount of variability as measured by standard deviations and coefficients of variations. By this technique three clear social types emerged. The first was the promiscuously social child who makes frequent contacts, both physical and verbal, with anyone available to him. The second was the selectively social child who limits his contacts but may be very talkative. The third was the nonsocial child who initiates few contacts and has a narrow range of responses in the time-sampling categories.

Correlations — The indices of frequency and range of verbal contact were positively correlated, .76+ or −.06 for boys and .54+ or −.10 for girls. There was an inverse relationship between the average frequency of work and the average frequency of speech found among kindergarten boys. This was not true of the girls.

Follow-up studies — These yielded definite though limited evidence of a considerable degree of constancy in the individual social pattern. In the case of two children who showed marked alterations in pattern, the changes were consistent with trends already noticeable in the previous pattern.

Author's interpretations

Much of this paper is devoted to a careful and detailed evaluation of the reliability of a time-sampling technique. The authors feel that the recording technique was less satisfactory than they had anticipated but, nonetheless, in the main quite reliable.

15

Ayer, M. E., and Bernreuter, R. G. A Study of the Relationship between Discipline and Personality Traits in Little Children, *Pedagogical Seminary and Journal of Genetic Psychology,* 50:165–170 (1937)

Setting: Nursery school

Subjects: 40 children

Time span: 1 semester

Meth. of obs. and test.: Topical interviews with parents; Merrill-Palmer Scale; general observations

The personality traits of children attending an emergency nursery school were compared with those of children from comparatively well-to-do homes attending a college nursery school. The disciplinary techniques of the parents and the effects of these techniques on the children were discussed.

Findings

Personality traits — The college nursery school children had higher sociability ratings than those children from the emergency nursery school. Also, the children from the college nursery school seemed more able to face reality.

Parental discipline — It was the original impression of the authors that the poorer parents were more likely to respond with temper, use physical punishment and exactment of a penance. Toward the end of the study, however, there were some reservations about this conclusion. When specific techniques were used to meet difficult situations, similar techniques were used by both groups of parents — e.g., on the one hand, isolating the child, punishing the child, or letting the child experience the natural results of his actions; and, on the other hand, rewarding the child for doing the right thing rather than punishing him for doing the wrong thing.

Authors' interpretations

Physical punishment seemed to increase the child's dependence on adult affection and attention and to decrease his tendency to face reality. Permitting the child to experience the natural results of the acts promoted independence of adult attention and fostered the child's ability to face reality. Forcing a child to do penance for his misdeeds seemed in the main to make him less able to face reality. Rewarding the child for doing the right thing rather than emphasizing punishment seemed to develop the quality of sociability with other children. The authors feel that there was a correlation between the parents' behavior and the development of the child's personality. A great show of temper by a parent often invoked a similar outburst by the child. Children whose parents were irresponsible or inconsistent were more dependent, less able to face reality, and developed personalities which tended to be less sociable and in general less attractive.

16

BALDWIN, A. L. Changes in Parent Behavior during Pregnancy; an Experiment in Longitudinal Analysis, *Child Development*, 18:29–39 (1947)

Setting: Fels Research Institute and home

Subjects: 46 families

Time span: 2½ years

Meth. of obs. and test.: Parent behavior scales obtained in prepregnancy, pregnancy, and postpregnancy

The author introduces the technique of multivariate analysis as a method for the interpretation of longitudinal data. This technique was described by S. S. Wilks in a paper called "Sample Criteria for Testing Equality of Means, Equality of Variances, and Equality of Co-Variances in Normal Multivariate Distributions," *Annals of Mathematical Statistics*, 17:257–281, 1946. All of the terms used in the article have special meaning as indicated in the Fels Parent Rating Scale. Cf. A. L. Baldwin, J. Kalhorn, and F. H. Breese, *The Appraisal of Parent Behavior*, Psychological Monographs, vol. 63, no. 4 (American Psychological Association, Washington, D.C., 1948). This is a manual providing a description and validation of the Fels Parent Rating Scale technique and includes data on reliability, validity, and use of the scale, as well as information on the essential interpretive procedures.

Findings

Warmth variables — The data indicate that there is a definite change in warmth variables after the arrival of a new baby. The parents are less child-centered, feel less rapport with the older child or children, and express less acceptance, approval, and affection.

Family contacts — The data also indicate changes in the variables having to do with family contacts. It was noted, however, that such changes are more marked after the birth of the second child than after the birth of the third or fourth child. In general, the sociability of the family decreases and the family policy becomes less democratic and less effective. The parents devote less time to explaining the family policy. More restrictions are imposed on the older child or children and more severe penalties are enforced. Babying and coddling decline appreciably but protectiveness and solicitousness fall off only slightly.

Author's interpretations

In general, the addition of another child to the family tends to reduce the warmth of family emotions and the amount of contact between parents and the other children. Although more restrictions are placed on the elder siblings, family control is less effective. During pregnancy, the author noted that although understanding increased, activity and suggestions decreased.

17

Baldwin, A. L. Socialization and Parent-Child Relationship, *Child Development,* 19:127–136 (1948)

Setting: Nursery school and home

Subjects: 67 children in longitudinal study; age 4 years

Time span: Not stated

Meth. of obs. and test.: Fels Parent Rating Scale; Fels Child Behavior Rating Scale

Findings

Democracy vs. control at home — Children from democratic homes show high degrees of aggression, fearlessness, and cruelty. They tend to be more active and show more ability in planning their own activities and in leading others. On the other hand, children from controlled homes show low degrees of aggression, fearlessness, and planning ability. They also tend to be negative, quarrelsome, and disobedient. Children from homes rated low in democracy and high in control tend to be quiet, well behaved, nonresistant, socially unaggressive, and restricted in curiosity, originality, and fancifulness.

Active democratic homes vs. inactive democratic homes — The author found statistically important differences in nursery school behavior from these two types of homes. Children from the active democratic homes were more aggressive and more active in their attitudes toward their environments.

18

BALDWIN, A. L. The Effect of Home Environment on Nursery School Behavior, *Child Development,* 20:49–61 (1949)

Setting: Nursery school

Subjects: 56 children; age 3 to 5 years

Time span: Not stated

Meth. of obs. and test.: Fels Parent Rating Scale; Fels Child Behavior Rating Scale

The three syndromes of variables which seem to have the greatest effect on the parent-child relationship were selected for analysis.

Findings

Democracy — Far more behavior differences in school can be attributed to democracy in the home environment than to any other variable. Democratically raised children are more active and more extraverted (friendly as well as hostile) and are favored in their group. They rate high in intellectual curiosity, originality, and constructiveness.

Warmth — Most democratic homes are characterized by warmth which provides emotional support.

Indulgence — This variable generally produces the opposite effects of democracy. It seems to be specifically related to physical apprehension and lack of skill in motor activities.

19

BALDWIN, A. L., KALHORN, J., and BREESE, F. H. *Patterns of Parent Behavior* (Psychological Monographs, vol. 58, no. 3). American Psychological Association, Evanston, Ill., 1945

Setting: Fels Research Institute and home

Subjects: 125 families

Time span: Not stated

Meth. of obs. and test.: Gesell Developmental Schedule; Merrill-Palmer Scale; Stanford-Binet Intelligence Scale — Revised Stanford Forms L and M; Cornell-Coxe Performance Ability Scale; Wechsler-Bellevue Intelligence Scale; Fels Child Behavior Rating Scale; Fels Parent Rating Scale; Rorschach Test; Thematic Apperception Test; teachers' records; interviews; tests of special skills

SYNDROME ANALYSIS

In this study the authors used the syndrome analysis of Sanford, *et al.* All sets of variables with a minimum intercorrelation of .60 or higher were chosen. These were then selectively examined and combined. The 3 central syndromes are democracy, acceptance of the child, and indulgence; some minor syndromes are severity, nagging, intellectuality, hustling, and personal adjustment. These syndromes reflect underlying emotional attitudes or personality traits and constitute patterns of variables around which parent behavior may be organized. In terms of these syndromes, homes are rated high, low, or inconsistent. Thus with 3 major syndromes and 3 ratings for each syndrome there are 27 possible classifications. Of the homes rated, 75% fall into 7 groups which serve as a basis for the study. The 3 major syndromes correlate highly, and low ratings on each syndrome have exceptionally high correlations.

TYPES OF PARENT BEHAVIOR

These groups, which comprise the 7 basic groups described, are defined and each is accompanied by illustrative case histories and charts.

Findings

Rejectant — In these cases the vector tended to travel in one direction since there are fewer ways to reject than to accept a child. Of the observed parents, 25% were rejectant and so rated low on the 3 central syndromes. Rejectant behavior was divided into nonchalant-rejectant (inert, passive) and active-rejectant (dominant, hostile). Both types tended to be arbitrary, coercive, dictatorial, and maladjusted with resentment, conflict, and quarrels.

Casual — The group which was neither acceptant nor rejectant is called casual and is also composed of 2 types, casual-autocratic and casual-indulgent.

Acceptant — Acceptant parents are divided into indulgent, democratic, and democratic-indulgent types.

CULTURAL PATTERNS OF PARENT BEHAVIOR

The data indicated that patterns of parent behavior are related to the parents' education, family income, and father's occupation.

Findings

Cultural characteristics — The rejectant parents tended to come from a lower economic stratum, but there was no marked difference between the nonchalant-rejectant and the active-rejectant. A few farm families were nonchalant-rejectant but most fell into the casual groups. Not a single farm family was democratic. The parents in the casual-autocratic group had the lowest I. Q.'s. Those in the casual-indulgent group also had relatively low I. Q.'s but a somewhat higher economic status. The acceptant-indulgent group was characteristically middle-class, but middle-class parents were also found in other groups. Most of the democratic parents had had college educations, and college teachers were nearly all democratic.

Authors' interpretations

No correlation is found between education and autocracy. All poorly educated parents are not autocratic, and all well-educated parents are not democratic. A good educational background seems necessary but not sufficient to ensure a democratic home.

PARENT BEHAVIOR IN RELATION TO PERSONALITY DEVELOPMENT OF CHILD

Findings

Intellectual development — In general, children's I. Q.'s were closely related to parental I. Q.'s. Increases in the children's I. Q.'s were noted over a 3-year period only in the democratic groups, in which the greatest change occurs, and in the casual-indulgent group. The acceptant-democratic group, but not the democratic-indulgent group, scored high on the variables of originality, planfulness, patience, curiosity, and fancifulness. The active-rejectant and casual-indulgent groups all scored low on the variables of tenacity, curiosity, fancifulness, and especially originality.

Emotional development — While democratic parents usually produced mild and placid children, active-rejectant parents had highly emotional children with poor control. Children of active-rejectant parents rated low on cheerfulness but those of nonchalant-rejectant parents were significantly high. In the preschool period children of democratic parents displayed a lack of sociability while those of indulgent parents were highly sociable, both in friendly and quarrelsome aspects. Children of active-rejectant families were somewhat resistant to adults in the preschool age. However, in the school years children of democratic homes were more sociable whereas those from indulgent homes became less sociable. Actively rejected children tended to accentuate early resistance and display marked sibling hostility.

Authors' interpretations

"In summary, we have found that the children, selected on the basis of parental behavior, do show consistent uniformities. For some groups, the democratic, the indulgent, and the actively rejected, the patterns are clear cut; for the others, they are only suggestive. The existence of these patterns lends further support to the syndrome analysis of parent behavior and leads us to believe that the method is empirically fruitful."

20

BALDWIN, B. T., and STECHER, L. I. *The Psychology of the Pre-School Child.* Appleton, New York, 1924

Setting: Nursery school

Subjects: 105 children; age 2 to 6 years

Time span: 3 years

Meth. of obs. and test.: Anthropometric measurements; case histories; mental traits; daily logs; Stanford-Binet Intelligence Scale; Detroit Kindergarten Test; Goddard Formboard; Porteus Maze; Three hole; perforation; Montessori Frames; tracing path; picture memory, picture completion, and picture vocabulary; color card sorting; color discrimination; weight and number concept

This was an early attempt at longitudinal study by means of the objective experimental technique and methods of observation. The authors were helped to define and observe behavior by noting five social attitudes of children: treating a playmate as an object, assuming an adult attitude, seeking attention, imitating the behavior of others, and cooperating with the group. A daily log was kept in the preschool laboratory emphasizing such types of behavior but these data were not subjected to statistical interpretation. In addition there were several individual history sketches of particular children. Many of the findings are descriptive and cannot be summarized. However, a few of the specific results are described here.

Findings

Retest scores showed a definite improvement over the original scores, which seemed to indicate that the children benefited from their preschool laboratory experience. The girls in this sample had a slightly higher I.Q. than the boys, and, interestingly enough, at age 3 the taller children seemed to be more advanced mentally than the shorter. Correspondingly, mental growth curves tended to parallel physical growth.

21

Banham, K. M. The Development of Affectionate Behavior in Infancy, *Pedagogical Seminary and Journal of Genetic Psychology,* 76:283–289 (1950)
Setting: Home and clinic
Subjects: 900 infants; age 4 weeks to 2 years
Time span: Not stated
Meth. of obs. and test.: General observations

Findings

Affectionate behavior and social concepts — The child's affectionate behavior develops along with his social concepts of people in general, of specific persons including the self, and of familiar and unfamiliar adults and children. When first developing, affectionate behavior is sexless in nature and direction. The outward form of affectionate behavior changes as the child adapts to the surrounding social situation but it always has the qualities of approaching, protecting, and giving warmth toward an object or human being. Affectionate behavior tends to prolong, repeat, or enhance the agreeable social situation; hostile reactions, on the other hand, are usually explosive and temporary, effecting gross changes or removal of the offending situation.

Chronological pattern of development — An infant under 5 months of age usually expresses affectionate behavior toward anyone who approaches him in a friendly manner. From age 6 to 12 months the infant behaves affectionately only toward familiar persons; unfamiliar persons may win affection in varying lengths of time depending on the individual child and the attendant circumstances. The 1- to 2-year-old child displays affection toward inanimate objects and toward himself. From age 2 to 3 years the child's affectionate behavior is usually directed toward adults, occasionally to other children. Patting, biting, etc. are compromises between affection and jealous hostility. During the preschool period the child learns to laugh and to accept good-naturedly interference which previously would have brought tears of protest. He continues to show affection for others despite their interference with his own desires.

Author's interpretations

The author comments that sadism does not appear to be a universal trait or phase of infant behavior. She also feels that erogenous exploration is not a concomitant factor in the early development of affection. Since the great majority of children who were seen in adopted homes one or more years after placement had developed ordinary patterns of affectionate behavior despite a neglectful, deprived, or restricted infancy, the absence of the parents need not lead to disturbance if an adequate mother-surrogate is available.

22

BARRETT, H. E., and KOCH, H. L. The Effect of Nursery-School Training upon the Mental-Test Performance of a Group of Orphanage Children, *Pedagogical Seminary and Journal of Genetic Psychology*, 37:102–122 (1930)

Setting: Nursery school

Subjects: 27 orphaned children; controls without nursery school experience; age 3 to 5 years

Time span: 6 to 9 months

Meth. of obs. and test.: Merrill-Palmer Scale

Findings

The average I.Q. for children with nursery school experience was 91.71 at the start of the test period and 112.57 at the conclusion; the corresponding averages for the controls were 92.59 and 97.71. On test performance, orphaned children averaged below the general population. However, when retested after 6 months or more of nursery school training, the orphaned children rated above the normal average.

Authors' interpretations

The authors infer that nursery school activities in some way raised the general accomplishment level of the institutionalized children.

23

BARUCH, D. W. A Study of Reported Tensions in Interparental Relationships as Co-existent with Behavior Adjustment in Young Children, *Journal of Experimental Education,* 6:187–204 (1937)

Setting: Nursery school

Subjects: 33 families with preschool children

Time span: 2.3 semesters (average)

Meth. of obs. and test.: Records of preschool behavior problems; teachers' ratings; mothers' reports; open-ended interviews with parents

Findings

Interparental tensions related to child behavior — There were four central interparental tensions which had the most significant relation to poor behavior adjustment in the child. The existence of these tensions between parents was related to factors in their own childhood familial relations. Tensions in the mother's background were more significantly related to the child's development than were those in the father's background.

Sexual adjustment in marriage — Interparental difficulties in this area often reflected compensation for, or prolongation of, a situation in the parent's own family. Specific situations were the mother's early attachment or antagonism to her own father, the father's early attachment or antagonism to his own mother, and disharmony of the mother's parents. In general, there was a poorer marital sexual adjustment among individuals whose attitudes toward sex had been gained from unwholesome education or superstition and had not been replaced by less repressive ideas.

Ascendance-submission — Interparental tension in this area was related to compensation or prolongation of earlier status in the parent's own family, particularly to the father's antagonism to his own father.

Consideration and sympathy — A lack of expression by parents in these areas was related to the mother's early attachment or antagonism to her father and disharmony between her parents.

Child-rearing methods — Lack of cooperation between parents about child rearing was related to a compensation for, or prolongation of, early status in the parent's own family and, in particular, to early antagonism between the father and his own father.

Other tensions — These include extramarital relations, problems related to poor health, inability to discuss problems together, insufficient affection, lack of normal friendship relations, and difficulties with relatives.

24

Baruch, D. W., and Wilcox, J. A. A Study of Sex Differences in Preschool Children's Adjustment Coexistent with Interparental Tensions, *Pedagogical Seminary and Journal of Genetic Psychology*, 64:281–303 (1944)

Setting: Preschool nursery

Subjects: 76 families with preschool children

Time span: 2 to 8 semesters (average)

Meth. of obs. and test.: Kawin Check List; Macfarlane Scale; open-ended interviews with parents; teachers' ratings

Findings

Interparental tensions related to child adjustment — Child adjustment was significantly related to interparental problems concerning sexual adjustment, a lack of consideration, a lack of ability to discuss differences and find solutions, a lack of expressed affection, and ascendance-submission.

Interparental tensions unrelated to child adjustment — These included problems concerned with cooperation in child rearing, extramarital relations, health, friends, work, relatives, pleasure pursuits, finances, criticalness of partner, and differing tastes.

Child adjustment related to interparental tensions — Girls

seemed more affected by parental tensions than boys, except for interparental problems concerned with ascendance-submission. However, the difference between the sexes in the proportions of satisfactory adjustment coexistent with parental tension was not statistically significant. The behavior adjustment of boys was significantly related to interparental problems concerning sexual adjustment, ascendance-submission, and lack of consideration; the behavior adjustment for girls was most significantly related to interparental tensions concerning sexual adjustment, lack of consideration, and lack of expressed affection.

Authors' interpretations

The less free verbalization between parents concerning a particular tension, the greater the degree of that tension. Tension closely related to the affectional and ego values of the parents seemed most significant.

25

BAXTER, A. The Adjustment of Children to Foster Homes; Six Case Studies, *Smith College Studies in Social Work,* 7:191–232 (1937)

Setting: Children's Service Association in Milwaukee, Wisconsin
Subjects: 6 children; age 4 to 18 years
Time span: Variable
Meth. of obs. and test.: Social work case histories

Findings

Foster home placement as a problem — Placement itself was a definite emotional problem only in the one case where earlier maternal deprivation had distorted the child's attitude toward this new experience. In every case but that of the 4-year-old, children in foster homes tended to develop those attitudes already acquired in their own family relationships. On the other hand, there was a readjustment to former parents in terms of foster parents. In general, displacement of hostile attitudes from parents to foster parents interfered with foster home adjustment and often led to new placements. This hostility was difficult to handle for both the children and foster parents. Children tended to be more intimate with foster mothers than with foster fathers.

Problems of adolescent foster children — Four of the adolescent children, feeling cheated of family support, were faced with a real problem in an environment where economic and social independence would be difficult. They were greatly concerned about leaving parental care and showed attitudes of emotional and economic dependency.

26

BAYER, L., and BAYLEY, N. Growth and Maturation of a Girl with Idiopathic and Precocious Puberty, *Stanford Medical Bulletin,* 11:241–252 (1953)

Setting: Hospital

Subject: 1 girl

Time span: 10 years

Meth. of obs. and test.: Medical case histories and examinations; endocrine studies; neuropsychiatric evaluations including electroencephalograms; x-rays; projective tests

The relationship between this girl and her mother was unusually close because the young mother had been widowed prior to giving birth and, perhaps, because of the developmental abnormality of the child.

Findings

Developmental deviations — The onset of the menses occurred when the child was 7 months of age, and menstruation subsequently became as regular as in a normal adolescent. Between the ages of 7 and 10 the child had 6 to 8 epileptic attacks with loss of consciousness. Throughout the study period she suffered from a more-than-average number of somatic illnesses, which were either preceded or accompanied by psychosomatic symptoms. Behavioral deviations began almost as early as the developmental abnormalities. In tracing the development of this child the authors found that when the pubertal spurt was superimposed on normal childhood growth, the results fell somewhere between the expected curves as these would be related to chronological or skeletal age.

Treatment — The child received treatment mainly designed to correct the developmental and behavioral deviations. Diet and androgen therapy proved unsuccessful but anticonvulsive therapy was somewhat effective. Both the child and the mother received several months of psychotherapy.

27

Bayley, N. A Study of the Crying of Infants during Mental and Physical Tests, *Pedagogical Seminary and Journal of Genetic Psychology,* 40:306–329 (1932)

Setting: Examination room

Subjects: 61 infants; 31 boys, 30 girls who were part of an extensive longitudinal study

Time span: Approximately first year

Meth. of obs. and test.: Records of crying during examinations

Findings

Causes — The infant's crying is catalogued under 1 of 13 identified causes which include specific test situations, continued handling, fatigue at the end of the test, internal conditions, colic pain, sleepiness, hunger, strange places and persons, putting the child down, interference with play activities, partial discomfort, spoiled behavior (trying to gain a specific end), and adverse conditioning. The most common cause is the specific test situation which often involved restriction of movement. After 6 months the most prominent causes are the strange examining place, the strange persons involved in the examination, interference, and adverse conditioning.

Crying related to sex, age, and various measured variables — There are no definite differences between the sexes regarding the quality or quantity of crying, or factors involved in it. In general, the trend indicates that crying becomes less marked in the examination situation until the age of 4 months; a sharp increase then occurs, especially at age 6 months. Crying is most frequent at age

1 month and from 6 to 10 months of age. There are no correlations between crying and mental test scores, socio-economic status, birth order, or other measured variables.

Inadequacies of measuring only quantity of crying — Merely measuring the quantity of crying is inadequate because it is a reaction to many stimuli and thus what appears to be periodicity is the result of many factors. Crying may be at a high level at the age of 6 to 10 months because a strange situation is recognized and thus an experimental artifact has been introduced. Each infant tends to be crying or noncrying rather than simply cyclical; also there is a consistency in the things which will make a particular infant cry from time to time.

Author's interpretations

She interprets her findings in the light of a theory of Jones and Bridges that crying represents a failure to make an adequate adjustment to a situation as it is perceived by the infant.

28

Bayley, N. Mental Growth in Young Children (chap. II); Factors Influencing the Growth of Intelligence in Young Children (chap. III), in: *Thirty-ninth Yearbook: Intelligence, Its Nature and Nurture,* Part II (National Society for the Study of Education). Public School Publishing Co., Bloomington, Ill., 1940

Setting: Not stated

Subjects: 48 children; and 61 children

Time span: First 9 years of life (tested 4 times a year)

Meth. of obs. and test.: California Test of Mental Maturity; Stanford-Binet Intelligence Scale

Findings

Chapter II (48 children)

Body build and skeletal maturity were unrelated to intelligence; nor was nursery school experience a significant factor in test score gains. The influence of other environmental factors such as socio-economic status was moderate. The author concludes that the most important factors in determining mental growth rates are endogenous ones.

Chapter III (61 children)

Individual children showed variable rates of mental growth. The most likely hypothesis for this is that intelligence is fundamentally innate and composed of many factors some of which do not start functioning at birth. Mental organization itself changes with growth, most rapidly in the first two years.

29

BAYLEY, N. Mental Growth during First Three Years (chap. VI), in: R. Barker, J. Kounin, and H. Wright (eds.), *Child Behavior and Development*. McGraw-Hill, New York, 1943

Setting: Not stated

Subjects: 61 infants

Time span: Tested from 0 to 3 years; 1 observation per month (50 minutes to 2 hours)

Meth. of obs. and test.: California Test of Mental Maturity; Stanford-Binet Intelligence Scale

Findings

Developmental tests administered before the age of 18 months were useless in the prediction of school age abilities, but tests administered between the ages of 2 and 4 years were of value. No correlations were found between the results of the tests and the amounts of crying in the test situation. The parents' I.Q.'s were significantly related to the children's later development, but not to their early rate of development. The most reliable factor in the prediction of a child's eventual I.Q. was the parents' educational level. These data are shown graphically by means of intelligence growth curves.

30

BAYLEY, N., and JONES, H. E. Environmental Correlates of Development, in: W. Dennis (ed.), *Readings in Child Psychology*. Prentice-Hall, New York, 1951

Setting: Not stated
Subjects: 59 families with newborn infants; age 0 to 6 years
Time span: 6 years
Meth. of obs. and test.: Berkeley Social Rating Scale; California Test of Mental Maturity (first year); preschool scale; Stanford-Binet Intelligence Scale; motor tests

Findings

Correlations between mental test scores, socio-economic factors, and age — Socio-economic factors were intercorrelated with the developmental sequence generally at ratios of .5 to .8. However, before the age of 18 months all of the factors showed zero or slightly negative correlations. After the age of 18 months the coefficient increased somewhat and by 4 or 5 years of age it began to have a significant correlation. An interesting finding was that the highest correlation between the mother's education and the test scores was found at age 2, whereas the highest correlation between the father's education and the test scores was at age 5. In both instances the ratio was .5.

Correlations between motor test scores and socio-economic factors — Motor test scores showed less significant correlations to environmental factors; in general, they were significantly less than those of the mental test scores.

31

BEESON, F. A Study of Vocational Preferences of High School Students, *Vocational Guidance Quarterly,* 7:115–119 (1928)
Setting: High school in Colorado
Subjects: 42 students
Time span: 4 years
Meth. of obs. and test.: Annual questionnaires

Findings

Persistence of vocational preferences of students — Specific vocational interests persisted for 1 year in 66% of the children, for 2 years in 43%, for 3 years in 34%, and for 4 years in only

10%. One-third of the children gave different choices every single year the questionnaire was given.

Vocational opportunities for students — A comparison of these high school students with the vocational opportunities in the community in general showed that the desired vocation was definitely out of proportion to the statistical opportunities in the community.

Author's interpretations

A predictable amount of frustration of goals was inherent in this situation.

32

BENDER, L., and GRUGETT, A. E. A Follow-up Report on Children Who Had Atypical Sexual Experience, *American Journal of Orthopsychiatry,* 22:825–837 (1952)

Setting: Psychiatric service

Subjects: 29 children who had some overt sexual experience; age 4 to 12 years

Time span: Variable

Meth. of obs. and test.: Social work case histories; psychiatric case histories

The study comprised two groups of children with atypical sexual experience at the time of referral. The children of the first group had been sexually involved with adults, while those in the other group had problems related to a confusion of sexual identity.

SEXUAL INVOLVEMENT WITH ADULTS

At the time of referral there were 10 girls and 4 boys aged 5 to 12. These children, who were unusually attractive and personable, had a scattering of behavior problems and many had neglectful home environments. They are considered in three groups.

Findings

Children sexually involved with their parents — A 5-year-old girl and her 6-year-old brother, who had been sexually approached by their father, later made adequate adjustments. Another 6-year-old boy, who had had sexual relations with his mother and had spent 3 years in a mental hospital, also seemed to have made an adequate adjustment at the time of follow-up. A 10-year-old girl, who had been approached sexually by her father, was found to have a chronic psychosis with auditory hallucinations and paranoid hostility at the time of follow-up.

Children with parental deprivations at the time of their involvement — A 7-year-old girl who had been in an orphanage and various foster homes, who had had sexual relations with a janitor at 7 years, and who at 19 years became the mother of an illegitimate child eventually married and made an adequate adjustment. A 9-year-old girl who had had open sexual relations, including intercourse, with a married man has also made an adequate adjustment at the time of follow-up. A 12-year-old boy who had been involved in pederasty later entered military service and was killed in action; a 7-year-old boy formerly involved in pederasty had made adequate adjustment. Of 2 girls, aged 11 and 12, who had had some sexual involvement with adults, one was later found to be somewhat delinquent and the other adequately adjusted.

Children with low I. Q. — A 10-year-old girl who had had illegitimate relations was hospitalized because of an episode of paranoia but at the time of follow-up had returned to the community on institutional support. A girl aged 9 on referral was diagnosed as an "inadequate personality" at follow-up. Of 2 girls aged 11 on referral, one was grossly disturbed and one had become sexually promiscuous at follow-up.

Authors' interpretations

The follow-up study showed that sexual disturbances were symptomatic rather than focal. The subjects tended to change constructively and seemed to abandon their sexual preoccupation. Overt sexual behavior in childhood did not necessarily forecast subsequent specific adult sexual maladjustment. There were 3 exceptions whose psychotic symptomatology limited their maturation and adjustment. The study further indicated that overt sex play with an adult was a deflection of normally developing impulses and was responsive to social and clinical treatment.

CHILDREN WHO WERE CONFUSED ABOUT THEIR SEXUAL IDENTITY

At the time of referral to the original study, there were 15 children aged 4 to 12 years. They are considered in three groups.

Findings

Children with homosexual parents (the dominant homosexual parent being of the same sex as the child) — A girl aged 8 on referral had had no sexual expression at all at follow-up. A boy aged 11 on referral was later dishonorably discharged from the Navy. A boy aged 8 on referral was later jailed for robbery and was thought to be psychotic. A girl aged 11 on referral was subsequently diagnosed as a schizophrenic and her brother was found to be a practicing homosexual.

Children who identified with parents of the opposite sex (the parent of the same sex being absent, hated, or ineffectual) — A boy aged 6 on referral subsequently developed schizophrenia and committed suicide. A girl and a boy, both aged 11 at the time of the original study, were later placed in institutions. Of 2 brothers, aged 8 and 9 on referral, one was hospitalized during follow-up and the other made only a borderline adjustment. A boy aged 10 on referral was diagnosed as a "borderline defective" at follow-up.

Children who identified with a parent of the opposite sex (having been deprived of the other parent in infancy) — A boy aged 4 on referral had developed dementia praecox and was markedly effeminate at follow-up. Another boy, seen in the original study at the age of 10, made a questionable adjustment and was not doing very well at follow-up. Of 2 boys, aged 8 in the original study, the first was found to be definitely disturbed and the second was tentatively considered to have made fair adjustment. A boy aged 11 on referral had been diagnosed as a "psychopathic personality" by the time of follow-up.

Authors' interpretations

No member of the group developed homosexuality during the follow-up period, but the brother of one member became a practicing homosexual. The authors were therefore unable to infer any specific effects on adult behavior from parent-child relationships but noted marked evidence of emotional and social maladjustment. All but 1 subject had required further extensive psychological and/or social care and 9 subjects were later diagnosed as schizophrenics.

33

Bender, L., and Yarnell, H. An Observation Nursery, *American Journal of Psychiatry,* 97:1158–1174 (1941)

Setting: Hospital observation nursery

Subjects: 250 children, of whom 128 were mentally defective; under 6 years of age at the time of admission

Time span: Variable

Meth. of obs. and test.: Stanford-Binet Intelligence Scale; psychiatric case histories; general observations

The group of 250 children was divided and analyzed in smaller groups.

MENTALLY DEFECTIVE CHILDREN

Of the 128 children in this group, 40 had been living in foundling institutions or had been referred by the courts as neglected children.

Findings

Family background — High percentages of psychopathology and social pathology as well as mental deficiency were found in the families of these children.

I. Q. — The I. Q. range was relatively high; only 3 children were idiots and 10 were borderline cases. Serial I. Q.'s were available for 28 of the 40 cases. In about 50% of the 28 cases the I. Q. tended to remain the same or to rise a little after placement in state schools for defective children; in the remaining cases there was a slight drop in I. Q. as mental age tended to become stationary.

CHILDREN HAVING NORMAL INTELLIGENCE BUT BEHAVIOR PROBLEMS

Of the 92 children in this group, 50% had been living at home, 20% had been referred by the courts as neglected children, and

30% had been referred by foster homes or institutions because of behavior problems.

Findings

Family background — Little evidence of pathology was seen in the family settings of 28 of these children. However, marked pathology was often found in the mother who frequently proved to be a neurotic person, unhappily married and personally dissatisfied. Quarreling was always present in the home and separation from the parents was often a threat to the children. The mother often resented the child and attempted to conceal this by overprotection and preoccupation with his physical ills.

Cause for admission to the nursery group — Most frequently the cause for admission was some crisis in the emotional life of the mother; often this was not confided to the doctor.

Symptoms found in the children — These included sleep disturbances, head banging, thumbsucking, eating disturbances, aggression, and temper tantrums.

Effect of nursery school experience — As a result of the socializing influence of the nursery group, many of the children improved. They lost a great deal of their anxiety, their neurotic symptoms lessened, and they were returned to their homes.

CHILDREN WITH A LOW-AVERAGE I. Q. AS COMPARED WITH THE PRECEDING GROUP

The children in this group were admitted to the observation nursery from social agencies and child-care agencies and upon recommendation by the children's courts.

Findings

Foundling home background and related problems — The authors felt that severe problems existed for children who, during early infancy, were placed in foundling homes where attachments could not be developed. They did not enter into group play but clung to adults. They were unable to accept love, exhibited temper tantrums when cooperation was expected, and were hyperkinetic and distractable. The authors classified these children as psychopathic personalities.

Effect of nursery school experience and good foster homes —

This group responded extremely well to the socialization process of the observation nursery and showed a rise in I. Q. When the children were placed in good foster homes, there was some rise in I. Q.

Authors' interpretations

"[We feel] that the infant cannot be raised in an institution without risking his normal personality development; we have learned that the only safeguard for the normal development of a child is a unified and continuous home environment for the first several years. The lack of this can never be compensated for."

34

BERNHARDT, K. S., MILLICHAMP, D. A., CHARLES, M. W., and McFARLAND, M. P. *An Analysis of the Social Contacts of Preschool Children with the Aid of Motion Pictures* (Child Development Series, no. 10). University of Toronto, Toronto, 1937

Setting: Nursery school

Subjects: 8 boys, 7 girls; age 22 to 27 months (Group I), age 29 to 30 months (Group II), age 37 to 42 months (Group III)

Time span: 4 or 5 months

Meth. of obs. and test.: Time samplings of social behavior; motion pictures taken of each child during complex social interaction

Findings

Social contact related to age increase — Few differences were noted in the frequency of social contact in the three age groups. However, increasing age was accompanied by increasing verbalization and concomitant complexity in social patterns. As age increased fewer changes or shifts were made with respect to objects or social contacts, and there were decreases in watching, with a correlation of −.7; in physical contact, with a correlation of −.38; and in approach and withdrawal, with a correlation of −.64. Adjacent use of material and uncooperative behavior also correlated negatively with age. Time and frequency of gestures and the cooperative use of material were traits that showed little differentiation in relation to age.

Comparison of the three age groups — During the 4-month observation period, the proportion of physical contact and adjacent use of material decreased in all groups. Verbalization and cooperative use of material increased in all groups. However, the changes were most marked in the 2 younger groups.

Types of social contact — There were 3 main groups which emerged: the first was the "straight-line" type in which the contacts of 1 child were equally distributed among all 4 members of the group; the "single-peak" type in which the child's contacts were largely confined to a single member of the group; and the "double-peak" type in which the child had a high frequency of contact with two other children. There were, of course, combinations of all 3 types, and all were affected by the increasing complexity of contacts which comes with age. A comparison of the profiles was of sociometric value. For example, the presence or absence of a reciprocal relationship (i.e., "reciprocal profiles" in which two children showed a mutual preference for each other) was of interest.

Authors' interpretations

The authors state that the use of motion pictures augmented their data by about 70%. In the main, the data on group behavior obtained by direct observation were merely confirmed by motion pictures, but the data on individual subjects were frequently changed. It is chiefly for the latter purpose that motion pictures are recommended as an observational technique.

35

BETKE, SISTER MARY ANGELA *Defective Moral Reasoning and Delinquency; a Psychological Study* (Studies in Psychology and Psychiatry, vol. 6, no. 4). Catholic University of America, Washington, D. C., 1944

Setting: Reformatory and school

Subjects: 50 reform school children and controls matched for I.Q.; age 11 to 15 years
Time span: Not stated
Meth. of obs. and test.: Social work case histories; special test of moral reasoning; McGrath Moral Information Test

The author attempted to find out whether any "pathological moral premises" were harbored in the subconscious of problem children.

Findings

Both the delinquents and controls had an equal amount of ethical knowledge. However, there was a definite difference in moral reasoning. Among the delinquents ethical factors were much less significant in moral reasoning, emotional reasoning was present in a slightly higher degree, and pragmatic reasoning in a significantly higher degree.

Author's interpretations

Although delinquent children knew right from wrong they were unable to apply this knowledge in moral reasoning with the result that their decisions were more emotional and pragmatic than those made by the controls.

36

BINNING, G. Health in the School [a series of articles based on a study of 800 school children in Saskatoon, Canada], *Health,* March–April, 1948; July–August, 1949
Setting: School
Subjects: 800 children
Time span: Variable
Meth. of obs. and test.: Wetzel Grid Studies of Growth; records of disease and trauma in the family; follow-up data

The observations are based on evidence from Wetzel Grid made during the school career of the subjects. There are no statistics presented in these papers. The findings are presented as two exact quotations.

Findings

Summarizing the 1948 article — "We found that events in the child's life that caused separation from one or both parents, death, divorce, and enlistment of a parent and a mental environment which gave the child a feeling that normal love and affection were lacking did far more damage to growth than did disease."

Summarizing the 1949 article — "Firstly, that every juvenile delinquent shows growth failure usually several years before delinquency begins. Secondly, with every adult chronic delinquent whose childhood growth records we have plotted on the Wetzel Grid, there has been such a deficit of growth when the child left us that permanent growth lag must have occurred. Thirdly, every suicide or attempted adult suicide has serious childhood growth disturbances usually not recovered from. Fourthly, the same phenomena are seen so far in every former student of whom we have growth records who later had to be admitted for mental disease."

37

BLANCHARD, P., and PAYNTER, R. H., Jr. The Problem Child, *Mental Hygiene,* 8:26–54 (1924)

Setting: Child guidance clinic

Subjects: 500 problem children; 337 unselected controls; age 4 to 16 years

Time span: Variable

Meth. of obs. and test.: Social work case histories; medical, psychiatric, and psychological examinations

When compared, the ratios of both white to colored and foreign-born to American-born parents were equal. There was, therefore, no tendency for racial or national distinctions in this series of 500 children. Seven fairly detailed case studies are included in this paper.

Findings

Differences between the problem children and the controls — Among the problem children there were 5 times as many mental defectives, 3 times as many individuals with personality disturbances, 6 times as many with conduct disorders, 3 times as many with marked physical defects, 2 times as many with speech defects, and 2 times as many cases with endocrine disturbance. Moreover, there were 2½ times as many problem children in grades to which their intellectual levels were unsuited even though there was very little grade acceleration and marked age-grade retardation in the problem group.

Factors associated with 250 of the problem children — In order of frequency these were bad home conditions, personality difficulties, poor heredity, poor physical condition, mental retardation, mental defects, endocrine disturbance, early illness, irregular school attendance and frequent school changes, and finally, emotional conflict.

Authors' interpretations

The authors derived the following three points from the individual case histories. The same problems might be attributable to various causes. Treatment was necessarily complicated and required medical, psychiatric, social, pedagogical, and vocational cooperation. Finally, the adjustment of the problem child is usually impossible without the cooperation of the home and school.

38

Blatz, W. E., and Bott, E. A. Studies in Mental Hygiene of Children; I — Behavior of Public School Children; a Description of Method, *Pedagogical Seminary*, 34:552–582 (1927)

Setting: School

Subjects: Number not stated; kindergarten through grade 8

Time span: 1½ years

Meth. of obs. and test.: Teachers' records of misdemeanors; psychiatric case histories

Misdemeanor is defined as behavior which interrupts teaching routine.

Findings

Misdemeanors — The number of misdemeanors reported for boys was greater than that reported for girls. Restlessness was the most common misdemeanor and then followed lack of application, disobedience, disorder, irregularity, uncleanliness, deceitfulness, showing off, emotional outbreaks, stealing, swearing, and timidity. The frequency of misdemeanors was not closely related to chronological age, except for the peaks between ages 8 and 9, and 13 and 14 in boys. The data for the latter age group suggests a preadolescent peak as mentioned by Haggerty. Also the frequency of misdemeanors varied inversely with I.Q. in boys, but not in girls. Only children had the best record; 62% of them committed no misdemeanors; none committed more than 16.

Prognostic importance of time of incidence in regard to miscellaneous data — Stealing in Sr. I is not as serious as in Sr. III. Uncleanliness is more serious in Sr. IV than in Jr. III. An emotional outbreak is more serious in the upper forms than in the lower. Deceitfulness is most serious when it appears after Sr. III. Disobedience is to be expected frequently in Sr. I; it should be considered more seriously in Sr. IV.

Authors' interpretations

Two years' experience with this method indicates that it offers a feasible means of recording behavior objectively and that it guards against research-staff prejudices and teachers' preconceptions.

39

Blatz, W. E., *et al.* *Collected Studies on the Dionne Quintuplets* (Child Development Series, nos. 11–16). University of Toronto, Toronto, 1937

Setting: School for child study

Subjects: Dionne quintuplets

Time span: Not stated

Meth. of obs. and test.: Anthropometric measurements; intelligence tests

Observations on the quintuplets were made by 2 trained observers from the age of 1 to 3 years; each child was compared separately with each of her 4 sisters.

Findings

Biological study — A complete biological study of the quintuplets was made in which fingerprints, palm prints, sole prints, and other features were compared as though there were ten pairs of twins. On the basis of this study it was assumed that the quintuplets were derived from a single ovum.

Mental growth — The apparent intelligence retardation of all the quintuplets might have been accounted for by three possible factors — prematurity, environment, and inheritance. Verbal skill was slowest and it was evident that gestures for intercommunication among the quintuplets were adequate for a long time. The general slope of all the quintuplets' development was the same; however, mental development itself was not identical.

Early social development — A large proportion of the time was spent watching each other. Each child differed qualitatively in social profile from the other, but individual trends of social behavior were relatively stable. The differences were thought greater than one would expect on the basis of their origin from a single ovum. The social contacts were separated into initiated contacts, responses to contacts, and initiated and response from contacts. The total frequency of all social contacts increased regularly with age. Social behavior trends were relatively stable.

Development of self-discipline — Individual differences were marked. The authors concluded that in this area influences on personality were environmental rather than hereditary.

Language development — The quintuplets were markedly slow in language development when compared with a control group of 13 single children. The authors suggested that speech development was slowed because of factors similar to those that operate in twins, who usually talk later than single children. In the case of the quintuplets these speech factors were magnified.

40

BLATZ, W. E., CHANT, S. N. F., and SALTER, M. D. *Emotional Episodes in the Child of School Age* (Child Development Series, no. 9). University of Toronto, Toronto, 1937

Setting: School and camp
Subjects: 3 groups — public school group, retarded group, and summer camp group
Time span: 3 months
Meth. of obs. and test.: Records of conduct; general observations

Findings

As age and intellectual quotient increase, emotional upsets at school decrease.

Authors' interpretations

The decrease of emotional upset at school is attributed to the development of alternative procedures. As intellectual manipulations become available to children they may replace emotional outbursts.

41

Blatz, W. E., and Griffin, J. D. M. *An Evaluation of the Case Histories of a Group of Pre-School Children* (Child Development Series, no. 6). University of Toronto, Toronto, 1936
Setting: School for child study
Subjects: 60 preschool children with developmental problems; 60 preschool children with no apparent difficulty
Time span: Not stated
Meth. of obs. and test.: Case histories

Findings

Differences between the problem children and the controls — The most common differences were faulty home discipline, poor adjustment beween self-assertion and self-negation, sleeping difficulties, and faulty bladder training. These were found more often in the problem group. Some of the areas in which no differences were found between the two groups were thumbsucking, nail-biting, other habit tics, and delayed speech.

Environmental factors — Poor discipline, inadequate accommodations, etc. were more frequently found, both absolutely and relatively, in the environments of the problem children than in those of the controls. The ratio between positive environmental categories and symptomatic categories was also higher.

Statistical analysis — Multiple-factor analysis of tetrachoric correlations in the case histories yielded two statistically significant factors. One factor was especially prominent in the environmental categories related to the social, practical, and disciplinary inadequacy of the home. It was also present in these symptomatic categories related to emotional difficulties and inadequacies of the self. Another factor was present in the categories related to child training such as difficulties in play and elimination.

42

BLUM, G. S., and MILLER, D. R. Exploring the Psychoanalytic Theory of the "Oral Character"; a Follow-up Study, *American Psychologist*, 6:339–340 (1951)

Setting: School

Subjects: 26 children in grade 4

Time span: Not stated

Meth. of obs. and test.: Time-sampling observations; sociometrics; teacher's rating scales; special experimental test

The original study, of which this is a follow-up, involved observations of objective oral activities such as mouth movements. These activities were significantly correlated with certain personality features.

Findings

Mouth movements continuing from original to follow-up study — The observational analysis of nonpurposive mouth movements in varied settings continued to reveal a fair measure of individual consistency, e.g., those who had a certain amount of nonpurposive mouth movement in the original study continued to have a similar amount in the follow-up. This study supported the finding that subjects with a large amount of nonpurposive mouth movements tended to have certain personality features including social isolation, lack of generosity, and a need for liking and approval.

Hypotheses introduced in the follow-up study — The inability to say "No," the desire for a magic helper, and a low frustration tolerance (none of which had been studied in the original paper) were introduced in the follow-up and showed no correlation with mouth movements.

43

BOTT, H. M. *Personality Development in Young Children* (Child Development Series, no. 2). University of Toronto, Toronto, 1934

Setting: School for child study

Subjects: 28 children; age 2 years 3 months to 4 years 11 months

Time span: 2 months

Meth. of obs. and test.: Diary and time-sampling techniques; measurements of variables were made as presented below

Children were roughly ranked and no real attempt was made to define the causes of acceleration or retardation.

SOCIAL AND MATERIAL ACTIVITIES

Findings

Type of material activity — The time in minutes spent in various types of material activity is scored for three age groups for a total of 150 minutes.

Age (in years)	*Type of material activity*		
	Idle or aimless	*Routine*	*Imaginative or constructive*
2 to 3	50.75	90	9.25
3 to 4	36	102.58	11.42
4 to 5	36	78.25	35.75

Use of material — The percentage of time spent in various types of activity is scored for three age groups.

Age (in years)	*Percentage of time*			
	Constructive	*Gymnastic*	*Locomotor*	*Manipulative*
2 to 3	3.8	19.5	34.7	42.0
3 to 4	7.2	17.9	37.5	37.4
4 to 5	13.9	21.4	39.6	25.1

Type of social participation — The time in minutes spent in various types of social participation is scored for three age groups for a total of 150 minutes.

Age (in years)	*Type of social participation*					
			Play with			
	Alone	*Watching*	*1 child*	*2 children*	*More than 2 children*	*Relations with adults*
2 to 3	34.63	29.37	58.75	13.50	2.13	11.62
3 to 4	29.25	19.75	53.92	20.91	7.92	18.25
4 to 5	16.88	5.75	50.12	33.63	19.25	24.37

Age in relation to behavior and types of activities — Neither the routine use of material nor play with one child is related to chronological age. Social behavior increases with age; adult relations show only a moderate increase with age. Negative correlations are found between age and idle or aimless behavior, watching, and playing alone, but they are too low for statistical reliability. Fairly high positive correlations are found between age and imaginative and constructive activity, play with 2 or more children, and relations with adults. Routine activity reaches a peak at age 3. Constructive play increases slightly from age 2 to 3 years, sharply from age 3 to 4 years. Idle behavior seems large for the 4-year-old children.

Relations among types of material and social activities — Every child participated in routine play as a form of material activity and in play with one child as a form of social participation. Aimless behavior has a positive correlation with playing alone and watching. Routine behavior has a 0 correlation with all social measures. A high positive correlation is found between constructive behavior and play with 2 or more children. Little correlation is seen between relations with adults and any other measure. Play with 2 children is the most representative single measure of social development. No significant relation is found between any of these behavior forms and I.Q. or mental development.

Author's interpretations

Social and mental development of normal children in preschool years is not parallel. Since idle behavior seems to be characteristic of 4-year-olds, the lack of sufficiently challenging material environment is indicated.

VERBAL AND MOTOR ACTIVITIES

Findings

Verbal activity — Individual variation in verbalization shows wide extremes. In 150 minutes of observation one child spoke 6 times; another spoke 753 times. Verbal activity shows a consistent increase with age and a positive correlation is found between total verbal behavior and chronological age.

Motor activity — Playing alone and passive motor behavior decrease with age. Motor activity with other children and with adults shows slight increases with age. No relation is seen between chronological age and motor activity. A study of the balance between self and social activities shows a preponderance of nonsocial over social forms of motor activity but this proportion decreases with age.

Relations among forms of verbal and motor activities — The most discriminating measures of development are passive motor development, talking to other children, and social motor behavior. Language reflects a social influence much more than motor behavior does. Commands, requests, and criticisms consistently increase with age. Responses reach a peak at age 2 to 3 years. Random verbalization and criticism consistently decrease. The expected parallel between verbal and motor behavior holds only in social categories. Correlations between types of verbal and motor activity and I.Q. give insignificant coefficients.

Comparison with other studies (Piaget, McCarthy, Day, Adams, Rugg) — Commands seem to have value as a predictive measure for social development. Children with high scores for commands seem to have the best social adjustment; children with low scores for commands seem to have the worst social adjustment.

SOCIAL RELATIONS AMONG CHILDREN

Findings

Balance of individual social relations — Social relations show a stable balance of "give and take." A child who is active toward others in both verbal and motor behavior receives most from them; a child who is not active toward others receives the least from them. However, a passive motor behavior trend is seen when a child who watches least is most watched by others.

Companionship and group social relations — Line pictures of interrelationships show various group configurations. Ten reciprocal relationships are distributed as 4-way friendships between the sexes; 3 friendships among boys, 1 friendship among girls, and 6 between girls and boys. Factors suggested as the cause of these relationships are age, friendship outside of school, and sibling relationships. In both verbal and motor behavior every age group is most active within itself, least active to those farthest away in age; this is a group trend, not a rule for individuals. A surprising diffusion of attention among all members of the group is seen for all children although some narrowing of the social range is observed as age increased. This narrowed social range shows evidence of some selectivity. The preference scale of children chosen as companions ranged from 0 to 20.

ADULT-CHILD RELATIONS

A second set of data is collected in addition to original records. A scheme of adult-child relations is formulated including an analysis of conversation, replies, responses, requests, stimulation, and imitation.

Findings

Of the total recorded behavior, 43% was verbal. No clear indication of age trends is observed. Individual differences in behavior toward adults are more important than in other aspects of social behavior. High and low scores may each have a great variety of interpretation; usually a low score is more desirable. In most cases, behavior to and from adults shows a close correspondence.

PERSONAL ACTIVITIES

Findings

These activities include talking to self, playing alone, tics, laughter, smiling, crying, repetition, and accidents. No relation is observed between these activities and chronological age. With some children talking to oneself is more or less persistent and it is not known if this indicates an introverted trend. Tics seem to be related to idleness or inactivity and to a lack of satisfying social adjustment. Both laughing and smiling are very prevalent among the 2-year-olds and the high-scoring children are also high in social participation scores. Laughing and smiling are not universal social mechanisms like speech, and they are used in varying degrees by different individuals. Verbal and motor repetition are observed to be *not* personal traits but part of verbal and motor tendencies in general and they do not measure adjustment. A .14 correlation between accidents and chronological age is observed. Accidents have a variety of observed causes.

44

BRACKETT, C. W. Laughing and Crying of Preschool Children, *Journal of Experimental Education,* 2:119–126 (1933)

Setting: Nursery school

Subjects: 29 children; age 18 to 48 months

Time span: Not stated

Meth. of obs. and test.: Time-sampling observations during free play

Findings

Laughter — Laughter even of nursery school children is a predominantly social form of behavior heard most frequently when children are interacting. Children who laugh often seem to prefer others who also laugh a great deal. Most of the children confine their laughter to a few particular children. The social contacts of the younger children are associated with laughter involving one other person; with older children at least two individuals are usually involved. Behavior remains consistent for a 1-year period. Individual children who laugh a great deal at the start of the year continue to do so at the end of the year; individual children who do not laugh a great deal also remain consistent for a year. As an overall pattern for the whole group, as age increases laughter increases.

Crying — Although the social situation is less well-defined than in laughter, crying also seems to be a predominantly social form of behavior. Contrary to laughter, no special relationship was observed between children who tend to cry and such children do not seem to prefer each other's company. Children who prefer each other seldom cry together. Crying occurs much less in routine or structured situations than in free play. Thus the author inferred that her speculation that children cry more when their activity is restricted was not substantiated. No relation is observed between the amount of language used and the frequency of crying; therefore crying was not thought to be a language substitute. Children who tend to cry a great deal have more physical contact and make less use of material than children who laugh a great deal. For the whole group, crying decreases as age increases.

45

Bridges, J. W. A Study of a Group of Delinquent Girls, *Pedagogical Seminary and Journal of Genetic Psychology,* 34:187–204 (1927)

Setting: School

Subjects: 33 girls; age 13 to 20 years

Time span: Variable

Meth. of obs. and test.: Social work case histories; National Intelligence Test; Myers Mental Measure Test; Mathews' Questionnaire for Emotional Stability in Children; Woodworth Test of Emotional Instability; reports from the institutions

Findings

Environmental factors — The histories revealed unfavorable home environments, with 70% of the girls coming from broken homes.

Intelligence — Although the range of abilities was considerable, on the whole, general intelligence (scholastic aptitude) was low. The group averaged 4 years of educational retardation. The M. A. average was 12 and the I. Q. average .85.

Temperamental factors — The delinquent girls were more unstable than the ordinary girls, especially in early adolescence. However, the differences between the two groups were less marked than those between comparable groups of boys. In regard to the amount of instability and its relation to age, delinquent girls were more like delinquent boys than ordinary girls were like ordinary boys. Delinquent girls showed at an earlier age the instability girls generally show in later adolescence; delinquent boys retained in later adolescence the instability boys generally show at an earlier age.

Delinquent girls compared with a group of college girls — More delinquent girls reported physical symptoms, insomnia, unhappiness, and abnormal impulses and conflicts, whereas more college girls reported unsociability, bashfulness, fear of responsibility, and worry over trifles. However, in the opinion of the observer holding a position of authority in the institution, the delinquent girls were more bashful and less subject to depression than they thought they were. Bright girls gave as many symptomatic responses as dull girls.

46

BRIDGES, K.M.B. A Genetic Theory of the Emotions, *Pedagogical Seminary and Journal of Genetic Psychology*, 37:514–527 (1930)

Setting: Nursery school

Subjects: 50 children

Time span: 3 years

Meth. of obs. and test.: Records of emotional behavior; daily observations

Author's interpretations

According to the genetic theory described in this article, all emotions derive from undifferentiated excitement which is produced by loosely coordinated visceral reactions and skeletal responses to any gross stimulation. Some of these visceral reactions eventually become differentiated and conditioned through certain stimuli. The various emotions are formed in early infancy when segments of visceral reactions combine with particular skeletal responses. Distress and delight are the first emotions to be differentiated from excitement. Distress is manifested by crying, tension, and checked breathing. Delight is manifested by smiles, cooing, and relaxation. Distress soon differentiates into components which, in combination with instinctive avoidance and aggression, form the emotions of fear and anger. Delight differentiates into joy (which is experienced in relation to events and objects) and affection (which is experienced in relation to persons). At nursery school age, specific emotions are distinguishable but the less differentiated emotions also continue to be observable at times. During childhood and adolescence, shame, jealousy, anxiety, elation, and parental and sexual affection evolve and become conspicuous. A particular kind of emotion can be identified more by overt behavior arising from a given situation than by accompanying visceral reaction, since the same emotion produces different visceral reactions in different people. Emotional behavior becomes more organized and modified as development proceeds. In terms of this scheme, emotions are largely acquired. Therefore, no two individuals are alike in the specific quality and quantity of their emotions because of constitutional and environmental differences. However, emotional reactions are somewhat similar for everyone because all individuals go through certain common experiences.

47

Bridges, K. M. B. *The Social and Emotional Development of the Pre-School Child*. Routledge, London, 1931

Setting: Nursery school

Subjects: 50 children; age 2 to 5 years

Time span: 3 years (for 9 months each year)

Meth. of obs. and test.: Records of emotional or social behavior; rating scales; daily observations

Weekly summaries of daily observations of social and emotional behavior of children in the nursery school provided the data for the main text of the book. No supplemental observations were made in the home. The observers attempted to make exact records of behavior rather than theoretical or categorical generalizations. Scales of behavior compiled from the data were applied. Cross-sectional descriptions of child behavior at various ages were compared. A character rating chart was also formulated and applied. The book concludes with studies of individual subjects. A central theme is indicated by the author's definition of emotional development as "the decreasing frequency of intense emotional responses, and the progressive transfer of responses to a series of stimuli determined by experience and social approval, and in the gradual change of the nature of the overt response in accordance with social dictates."

Findings

Individual studies — By means of a sequential, developmental scale six children were described in detail. The transition of a seclusive and inhibited emotional child into a normal, sociable and expressive child is described. Simultaneous with this transition, weight gain and a surge in physical development occurred. Another subject was an unsociable, cold, and "stuffed-up" child who became normal and sociable during the course of the study. A sociable, cheerful, and uninhibited child consistently remained that way. No change whatsoever was observed in an unsociable, distressed, and schizothymic child. Another subject, manic-depressive in behavior, was extremely sensitive to stimuli with an intense startle response. He was big for his age and sociable but overly possessive. The child had a characteristic posture of bent knees and open mouth. No significant changes were observed in his behavior during the study. An inconsistent disciplining technique in the home of another child was thought to be responsible for his manner of interacting. He was hyperactive, impulsive, cheerful, and sociable.

48

Bridges, K. M. B. Emotional Development in Early Infancy, *Child Development*, 3:324–341 (1932)

Setting: Foundling and baby hospital

Subjects: 32 infants; age 3 weeks to 2 years

Time span: 3 or 4 months

Meth. of obs. and test.: General observations; detailed records of emotional behavior

The author expounds her hypothesis that the hierarchy of emotional responses is derived from a single basic response (excitement), rather than three basic responses as claimed by the behaviorists. As the infant's emotional development proceeds, general responses differentiate into specific responses, which in turn combine to produce complex acts.

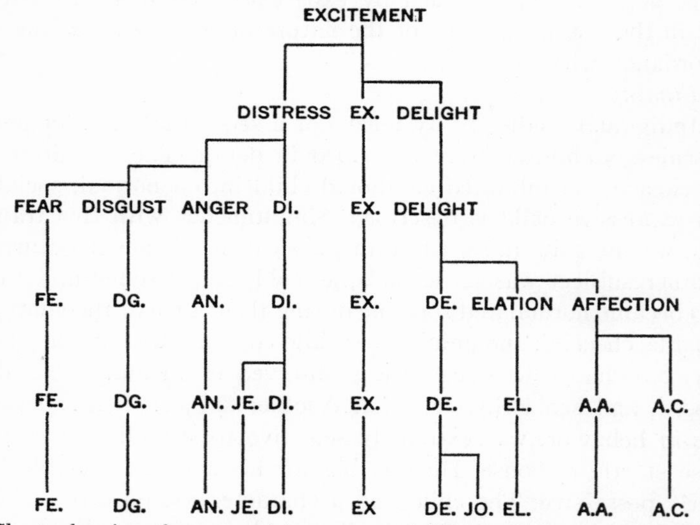

Chart showing the approximate ages of differentiation of the various emotions during the first two years of life. The age levels, beginning with Excitement at the top, are birth, 3 months, 6 months, 12 months, 18 months, and 24 months, respectively.

Key: A.A. — Affection for adults; A.C. — Affection for children; An. — Anger; De. — Delight; Dg. — Disgust; Di. — Distress; El. — Elation; Ex. — Excitement; Fe. — Fear; Je. — Jealousy; Jo. — Joy.

49

BRIDGES, K. M. B. A Study of Social Development in Early Infancy, *Child Development,* 4:36–52 (1933)
Setting: Foundling and baby hospital
Subjects: 62 infants; age 3 weeks to 2 years
Time span: 3 months
Meth. of obs. and test.: Records of behavior in social situation; general observations

Findings

Chronological trends of social development of child-adult relations — Adults receive the earliest social interest. Differentiation and recognition of individuals and objects develop later. Reciprocal, playful action and imitative vocalization appear as intermediate behavior forms. At age 4 to 5 months conflict develops between the desire for attention and the dislike of consequent interference with individual liberty. At age 7 to 12 months the infant welcomes and copies familiar persons and disapproves of the stranger with whom no pleasant associations exist. By age 1 year imitation and some disobedience are recognizable. About 6 months later a temporary period of obstinate reaction develops and then the child becomes cooperative again.

Chronological trends of social development of child-child relations — Interest in other children develops a few weeks later than interest in adults. This interest progresses slowly until age 9 months, when rapid development begins. Preferences are shown by age 14 months. At age 15 to 24 months aggression is common and "mob behavior" can be observed in the absence of a personal authority figure. In the presence of an accepted authority the child is fairly orderly, disciplined, and friendly.

50

BRIDGES, K. M. B. Occupational Interests of Three-Year-Old Children, *Pedagogical Seminary,* 27:415–423 (1934)
Setting: Nursery school
Subjects: 10 children

Time span: Not stated
Meth. of obs. and test.: I.Q.'s; Stutsman Performance Test; records of materials used; general observations

Findings

Variations of interest — In general there was considerable diversity in the children's interests. Cylinders, bridges, and color pairs were the materials most favored by both boys and girls. However, there were occasional wide individual variations. For example, throughout the study two boys showed a marked and consistent interest in building with large bricks.

Variations of interest between boys and girls — Girls showed more diversity in their preferences than boys. Boys seemed to prefer large building material, freedom of action, and activities involving gross movement and a possibility for originality. The girls seemed to enjoy activities with less movement, less need to discriminate color, size, etc., less finger manipulation, and less originality, i.e., they enjoyed sitting comfortably at a table and following instructions, and such activities as fitting graded cylinders into holes.

Occupations of least interest — Building a one-story tower, sweeping with brush and pan, placing material in a frame, and playing with a tea set or stuffed animal were of least interest to the children.

Time spent on occupation — The group as a whole had an occupation span of 8.1 minutes, for boys 7.3 minutes and for girls 8.5 minutes. The average for the longest time span on a single occupation was 37.5 minutes for boys and 34.5 minutes for girls.

Author's interpretations

The author indicates that familiarity, difficulty, and lack of sufficient creative value are plausible reasons for unpopularity of certain items, such as those mentioned above. The study was of value, the author feels, in planning and understanding the activity of a preschool group.

51

Bridges, K. M. B. The Development of Primary Drives in Infancy, *Child Development*, 7:40–56 (1936)

Setting: Foundling home

Subjects: 72 children

Time span: 4 months

Meth. of obs. and test.: Timed observations

The study is descriptive and categorical. No statistical observations are involved.

Findings

The infant exhibits a "drive for survival" from birth onward. At age 1 to 2 months curiosity, postural adjustment, and sociability begin to appear. At age 3 to 6 months the types of behavior that develop are acquisition, vocalization, assertion, shyness, retreat, and social submission. At age 6 to 12 months there is evidence of aggression, imitation, destructiveness, and possessiveness. At age 1 to 2 years the infant's behavior pattern includes locomotion, creative construction, vocal communication, obstinacy, a concept of revenge, flight and avoidance, and social compassion. Most of these forms of behavior are still evident at age 2 to 5 years and, in addition, social stimulation has begun to develop.

52

Brown, G. D. The Development of Diabetic Children, with Special Reference to Mental and Personality Comparisons, *Child Development*, 9:175–184 (1938)

Setting: Diabetes clinic

Subjects: 42 children; median age, 13 years

Time span: Not stated

Meth. of obs. and test.: Open-ended interviews with children and parents; questionnaires; medical histories and examinations; Stanford-Binet Intelligence Scale; school histories; Woodworth-Cady Psychoneurotic Inventory; Otis Group Intelligence Tests

The results of this study differed from Joslin's in that the

I. Q.'s of the diabetic children did not exceed those of their siblings. Although the Woodworth-Cody Psychoneurotic Inventory revealed no differences between the diabetics and their siblings, parents rated the diabetics more cautious, stubborn, excitable, and irritable.

53

Bruch, H. Obesity in Childhood and Personality Development, *American Journal of Orthopsychiatry*, 11:467–474 (1941)

Setting: Pediatrics department

Subjects: 200 children; 25–150% overweight

Time span: Variable

Meth. of obs. and test.: Pediatric examinations; case histories; psychiatric case histories

Findings

Child development factors — The children in this study showed accelerated physical and mental development but greatly retarded emotional and social maturation. For example, 40% of the children over 6 years of age were still enuretic. Obesity was attributed to excessive, preferential food intake and low motor activity.

Family factors — The family factors included a weak father, a domineering mother, and a lack of home acceptance of basic childhood needs. Often the mother, having experienced poverty and hunger in her own childhood, reacted to her deprivations and disappointments with self-pity, overprotectiveness of the child, and undue emphasis on food.

Author's interpretations

The obese child, unable to express himself dynamically, derives satisfaction exclusively from food and manifests his creative strivings through his large size. This largeness both makes an impression on his environment and protects him against a threatening outside world. Since the reaction of an obese child is a "somatic compensation," there is much resistance to treatment.

Therapy should attempt to provide more dynamic outlets for his creative drives.

54

Brunet, O., and Lezine, I. Psychologie de la première enfance; une contribution du groupe des jeunes parents, *Enfance,* 2:355–363 (1949)

Setting: L'Institute d'Orientation and L'Institute de Psychologie

Subjects: 12 infants

Time span: Not stated

Meth. of obs. and test.: Parents' diaries; motion pictures; developmental examinations

This article analyzes only the data pertinent to the child's reaction to his own mirage in the mirror. Language comprehension and the use of pronouns are discussed briefly.

Findings

At 2 months of age, 11 out of 12 infants focused on their reflection; by 4 months of age they smiled at it. [These findings are in disagreement with the data presented by R. Zazzo, *Enfance,* 1:80–85, 168–175, 365–372 (1948).]

55

Bühler, C. Der Pubertätsverlauf bei Knaben und Mädchen, *Zeitschrift für Sexualwissenschaft,* 14:6–10 (1927)

Setting: Institution

Subjects: 50 girls in prepubescent stage

Time span: 9 months

Meth. of obs. and test.: General observations

Findings

A negative phase occurs from 2 to 9 months prior to menarche. During this time girls isolate themselves. They are restless and mentally and physically uneasy. The phase has an abrupt end and normal social interaction recurs.

56

Bühler, C. Die ersten sozialen Verhaltungsweisen des Kindes, *Quellen und Studien zur Jugendkunde,* 5:1–102 (1927)

Setting: Home and institution

Subjects: 146 infants

Time span: Not stated

Meth. of obs. and test.: Developmental examinations; general observations

The author attempts to describe the earliest forms of social contact and the emergence of various types of personality — dominating, exhibiting, producing, and humanitarian. The discussion is supplemented by tables and individual records.

Findings

Contact was assumed to begin when one infant eyed another. This was followed shortly by sympathetic crying. By the age of 3 or 4 months infants began to smile at one another. By 6 or 7 months they began to treat one another as objects of play. During this period they also began to offer objects to one another and to grasp objects in another's possession. At the end of 12 months this often developed into cooperative play.

57

Bühler, C. An Observational Study of the First Year of Life, in: W. Dennis (ed.), *Readings in Child Psychology.* Prentice-Hall, New York, 1951

Setting: Home and institution

Subjects: 69 infants; 28 from private families, 41 from institutions

Time span: Not stated

Meth. of obs. and test.: Timed and general observations

Findings

Sleeping situation — The greatest decline in the hours of sleeping occurred during the first four months. By the fifth month waking time equaled the sleeping time. Dozing gradually decreased throughout the entire first year but there was a decisive decrease during the eighth month.

Negative and positive behavior — Positive reactions appeared much later than negative ones but during the first year became preponderant; the greatest increase in positive reaction occurred during the fifth and ninth months. Behavior expressing pleasure and satisfaction increased throughout the first year.

58

Burgum, M. Constructive Values Associated with Rejection, *American Journal of Orthopsychiatry,* 10:312–326 (1940)

Setting: Child guidance institute

Subjects: 25 children rejected by parents

Time span: Not stated

Meth. of obs. and test.: Psychiatric case histories; social work case histories

Findings

Of the 25 rejected children, 20 were described as independent, 15 were capable of self-amusement, and 13 had special interests or hobbies. A good adjustment outside the home was made by 10 children; 5 were called responsible; 8 responded to affective relationships; and 10 showed evidence of early maturity. There was some indication of a correlation between I. Q. and ability to make constructive use of rejection.

Author's interpretations

Independence may depend on several etiologic factors. It may be caused by defiance ("You don't love me, and I don't need you"); it may express aggression or revenge; it may be rewarding in itself. The rejected child is *obliged* to protect himself. He seeks to compensate for deprivation experienced at home. He looks for praise, rewarding experiences, and outward expression of his fantasy life. Readjustment of rejected children may be aided by exploration of the constructive elements in this experience. However, it should be remembered that independence based on reaction formation may not be durable.

59

BURKS, B. S. A Scale of Promise, and Its Application to Seventy-One Nine-Year-Old Gifted Children, *Pedagogical Seminary and Journal of Genetic Psychology,* 32:389–413 (1925)

Setting: Not stated

Subjects: 71 children with I. Q.'s of 140 or higher; 46 boys, 25 girls

Time span: Age 9 to adulthood

Meth. of obs. and test.: Rating scales

Findings

The mean standing of these gifted children was somewhat higher than average on each of ten items of a promise scale: persistence, moral standards, vocational interest, social adaptability, I. Q., health and physical energy, family social status, interest in activities, originality, and ability to profit from school work (boys only). The boys attained a slightly higher average score than the girls. In the sample studied, I. Q. and promise could not be exactly correlated. There was a wide variation in scores. Some of these gifted children attained scores which indicated less promise than average children, while others attained scores indicating future eminence.

60

BURLINGHAM, D. T. Twins; Observations of Environmental Influences on Their Development, in: *Psychoanalytic Study of the Child* (vol. II). International Universities Press, New York, 1947

Setting: Nursery

Subjects: 3 sets of twins

Time span: Not stated

Meth. of obs. and test.: General observations with diaries; detailed sleep charts; developmental charts

Author's interpretations

In the early stage of their development twins react to the mother just as single children do. Later they sense their mother's pleasure in each of them and in the comparisons she makes between them, thus becoming conscious of each other and their mother's pleasure in them as a unit. Eventually they meet this reaction from the rest of the people in their environment as well. They may have a greater desire for individual attention because they receive it less frequently than the single children, but they also experience the desire "to please à deux." Unless identical twins are separated, each may feel that he is unique for only one other person, the twin. In consequence they turn to each other for certain missing elements in the mother-child relationship and all other childhood relationships. The mother's problems and attitudes in regard to the twins is an important factor in determining their development. Narcissism, exhibitionistic tendencies, and guilt, if present, may prevent identification with the twins and may be reflected in the personality structure of the twins themselves.

61

BURLINGHAM, D. T. The Relationship of Twins to Each Other, in: *Psychoanalytic Study of the Child* (vol. III). International Universities Press, New York, 1949

Setting: Nursery

Subjects: 3 set of twins admitted in infancy

Time span: 2 to 4 years

Meth. of obs. and test.: General observations with diaries; detailed sleep charts; developmental charts

The sample used in this study was too small to permit statistical generalizations which would be valid for other twins. But the descriptive data are detailed and so afford many insights into these particular twins.

Findings

Chronological development — During the first 4 months the twins were in separate bassinets and took no interest in each other. From the age of 5 months all the babies showed growing attachment to their mothers and the nurses who cared for them. At the age of 8 months they responded to the playful overtures and speech of people they knew. At the age of 10 months they began to notice and compete for attention. More emotional reactions were shown in the feeding situation than in any other. Often the children of a pair behaved quite differently. The fact that dominant characteristics produce an active and a passive twin is easily observed at the toddler stage. The passive twins were not necessarily content with their subordinate roles. They became aware of differences in achievement at an early age, and of their mothers' pleasure as a reward for achievement. This fostered competition.

Imitation — Imitation is a normal phase of development, usually beginning at the age of 9 months. The twins began imitating somewhat later, at the age of 12 months. By the end of 13 months imitation had become a game involving fairly simple body movements. In one set of twins, as these movements grew more and more complex, the game seemed to hinder their mutual relationships and their ability to make normal contacts with other people. Another pair of twins resorted to imitation for dependency reasons. In many situations contagion of feeling was observed in all twins. In one pair each twin copied the other as a competitive technique to prevent the sibling from obtaining an advantage in the relationship, especially when strong emotions were involved.

62

BURLINGHAM, D. T. *Twins; a Study of Three Pairs of Identical Twins.* International Universities Press, New York, 1952

Setting: Nursery

Subjects: 3 sets of twins

Time span: Age 21 months to 4½ years (set A); age 4 to 14 months (set B); age 4 to 19 months (set C)

Meth. of obs. and test.: Detailed charts of development of habits; diaries kept by author; protracted observations

Findings

Twin interaction — Twins' interaction began at about the age of 8 months. Competition for attention developed early, especially in the feeding situation. Signs of activity and passivity developed soon after. Attempts at mutual imitation led to identical behavior. In one set of twins this was thought to hinder psychomotor development. Twin interaction seemed not only to provide security under disturbing conditions but also to serve as a barrier against new contacts.

Separation — When separated from each other, the twins sought comfort in words, in play, in the search for substitutes, and even in the use of mirror images. Defensive reactions included fears, mourning states, identification with the missing twin, and attempts at sublimation.

Jealousy and competition — Jealousy and rivalry between the children were definitely affected by the twin relationship. Strong positive feelings involving protection, sharing, and other forms of behavior suitable for overcoming jealousy and conflict were developed through mutual education. Competition for attention presented a special problem. One solution appeared to be division of time for individual contact. It was noted that the twin relationship was somewhat disturbed during separation from the mother and that the balance was re-established when contact was resumed. Sometimes there was exchange of roles in relation to the parent.

Fantasy and reality twins — The author made a comparison between a child seeking attention and companionship through a fantasy twin and a child having a real twin. Whereas the fantasy twin fulfills narcissistic needs, the real twin does not and he presents such problems as rivalry and jealousy. The fantasy twin can be relinquished but the real twin is a constant.

63

Burlingham, S. Therapeutic Effects of a Play Group for Pre-School Children, *American Journal of Orthopsychiatry,* 8:627–638 (1938)

Setting: Play group
Subjects: 10 preschool children
Time span: 2 years

Meth. of obs. and test.: Social work case histories; family records; home and school visits

Findings

Changes in the play group children — The play group was utilized constructively by six children, who retained their gained values at home and school and developed normal interpersonal relationships. However, a high degree of anxiety in the homes of two seriously disturbed children prevented them from obtaining much benefit; one of the children was a shy 3-year-old, the other a destructive child who suffered from maternal rejection and sibling rivalry.

Environmental changes — Noticeable improvements in the families' circumstances and in the parents' attitudes occurred in five instances. These constructive changes in the home situation were attributed partly to the play group program and partly to case-work service. Little modification was shown in three other homes.

Author's interpretations

The degree to which a child can utilize the growth opportunities offered by a play group depends largely on parental relations at home.

64

CAILLE, R. K. *Resistant Behavior of Preschool Children* (Child Development Monographs, no. 11). Teachers College, Columbia University, New York, 1933

Setting: Nursery school

Subjects: Number not stated; age 19 to 49 months

Time span: 1 year

Meth. of obs. and test.: Time-sampling observations; stenographic records of child's language over a 2-day period and during intelligence tests

Resistance, as used here, must be considered as the failure to comply with a suggestion, usually from an adult, or a direct expression of a refusal of the person's request by physical or vocal means.

Findings

Correlation of age and behavior — By every technique except language records, the peak for resistant behavior was noted at 3 years of age, plus or minus 2 months. The peak for acquiescent behavior was noted 4 months preceding the peak for resistant behavior. Numerically the 2-year-olds and 3-year-olds showed the same amount of acquiescent behavior, but the 3-year-olds showed twice as much resistant behavior as the 2-year-olds. As physical resistance decreased with age, vocal resistance took its place. The latter was highest among older children but had a secondary peak at age 3 to 3½ years.

Intelligence — No significant relation was observed between I.Q. and resistance, acquiescence, or aggression.

Frequency and technique — Definite individual differences were observed in the frequency of each type of behavior and in the kinds of resistance techniques.

65

CAMERON, W. J. A Study of Early Adolescent Personality, *Progressive Education,* 15:553–563 (1938)

Setting: An "extracurricular" clubhouse; grade 5

Subjects: 200 school children; age 9 to 11 years

Time span: 8 years

Meth. of obs. and test.: Not stated

The clubhouse where the study was undertaken provided opportunities for social experience and maturation. The children were considered to be fairly representative of a local school system. They were observed as individuals and as members of an interacting group. The problems and progress of group dynamics at this age level are discussed in a conversational manner.

66

CAMPBELL, E. H. The Effect of Nursery School Training upon the Later Food Habits of the Child, *Child Development,* 4:329–345 (1933)

Setting: Summer camp
Subjects: Experimental group (nursery school experience) — 11 boys, 7 girls; average age, 9 years 5 months; average I.Q., 125. Control group (no nursery school experience) — 8 boys, 7 girls; average age, 9 years 3 months; average I.Q., 125
Time span: 6 weeks follow-up of children previously in a longitudinal group; time intervals of follow-up varied from 1 to 9 years
Meth. of obs. and test.: Rating scales; records of eating behavior

Findings

There was little difference in the food habits of the two groups. Categories of frequency of faulty habits, use of devices to avoid eating disliked food, amount of food ingested, and time devoted to eating were about equal. The most common faulty habits were aversion to particular foods or ways of food preparation, reluctance to finish entire meal, interference with eating by overtalkativeness, overfondness of sweets and starches, and bolting food. There was some evidence that the nursery group ate more milk and vegetables, the controls more eggs and bread.

Author's interpretations

There is slight evidence for concluding that a long period of nursery school attendance (6½ semesters) does not guarantee good habits. However, there is some evidence that the more recent the nursery school experience, the better the food habits. The food habits of children from the same home are likely to be similar, even when one of a pair has attended nursery school and the other has not. The author concludes that the home is a much more important factor than the nursery school in forming food habits and attitudes.

67

CAMPBELL, E. H. The Social-Sex Development of Children, *Genetic Psychology Monographs,* 21:463–552 (1939)
Setting: Merrill-Palmer School clubs

Subjects: 112 graduates of nursery school; 59 boys, 53 girls; average age — 9.9 for boys, 9.57 for girls; average I.Q., 117
Time span: 3 years
Meth. of obs. and test.: Skeletal age; I.Q.

A list of 149 statements was compiled and subjects had repeated ratings. By assigning values in terms of the social development variables a quantitative determination of increasing social-sex behavior was obtained. This paper contains an excellent survey of the literature on the subject.

Findings

Correlations and grouping of social-sex development — A stepwise progression of social-sex behavior was found with earlier sequences for girls; age 10-14 years was the most variable period. Of the 34 children who were rated for 3 years, 12 remained in the middle quartile. Only 2 of the other 22 (who rated high or low) remained in the same position for 3 consecutive years. Social-sex development showed low but positive correlations with skeletal age, mental age, and I.Q. Three stages in social-sex development were noted: (1) little sex discrimination — age 5-9; (2) unisexual behavior preceding adolescence — age 9-14; (3) gradual preference for the opposite sex — age 14-17. At each stage, the girls' development was approximately 6 months ahead of the boys' development.

First stage — During the first stage the children were not self-conscious about bodies or physical contact. At age 6, for example, they ignored sex in the choice of playmates.

Second stage — Touching members of the opposite sex made girls self-conscious at 11 and boys at 12 years of age. Girls whispered incessantly at 12 and classified games according to sex at 13; they were obviously attracted to boys at 13½ years but would not admit it. Boys showed no interest in the opposite sex until age 14.

Third stage — Sex modesty began for both girls and boys at age 14. Girls began to primp at 14½ years and showed heterosexual interest more openly. Boys lost interest in adult attention at 15 years, but girls did not. The cultural role was taken up by girls at 15, by boys at 15½. Girls were interested in boys in general, but not in particular boys at 15. Boys did not reach this stage until 15½. However, the physical contact of dancing was enjoyed by boys at 15½ and by girls at 16. No examples of "crush" or hero worship were found.

Author's interpretations

The author compares the results of her study with those of other investigations and presents two case histories. She points out that her findings disagree with those of Parten and Koch who found earlier unisexual behavior.

68

CAMPBELL, E. H., and BRECKENBRIDGE, M. E. An Experiment in the Study of Individual Development, *Child Development,* 7:37–39 (1936)

Setting: Merrill-Palmer School
Subject: 1 girl
Time span: 20 months to 8½ years
Meth. of obs. and test.: Rating scale

This paper is a summary of a manuscript that fully described the method of longitudinal study at the Merrill-Palmer School. The authors describe one girl in terms of a constellation of traits, a life chart devised by the Merrill-Palmer School, and a series of graphs outlining different aspects of her growth and development.

69

CAMPBELL, R. V. D., and WELCH, A. A. Measures Which Characterize the Individual during the Development of Behavior in Early Life, *Child Development,* 12:217–240 (1941)

Setting: Not stated
Subjects: 41 children, including 5 sets of unisexual twins
Time span: 4 years
Meth. of obs. and test.: Developmental examinations

Findings

On the basis of behavior, the individual was poorly differentiated from the group at birth; in terms of rate of change in development, the individual was more precisely different from the group, but the differentiation was not sufficient to characterize the individual. Judging from behavior achievements, differentiation from the group increased up to 150 days, at which age a rank order of the 40 children showed the twins in adjacent positions in 4 out of 5 cases. On the average, twins developed more slowly than others in behavior activities. Girls were slightly more advanced at birth in development than boys.

Authors' interpretations

The progressive differentiation of the individual suggests development of the nervous system. It is unlikely that scales based on achievement in behavior can yield satisfactory appraisals of the individual under 3 months of age. Later such scales may provide increasingly accurate means of assessing individual attainment.

70

Carey-Trefzer, C. J. The Results of a Clinical Study of War-Damaged Children Who Attended the Child Guidance Clinic, *Journal of Mental Science,* 95:535–559 (1949)

Setting: Guidance clinics

Subjects: 212 children

Time span: Variable

Meth. of obs. and test.: Psychiatric case histories; psychiatric interviews

Findings

Air raids, evacuation, and changes in family life all contributed to disturbances which were apparent in four symptom complexes. Of the group 71.2% were more aggressive, 55.1% had fears and anxieties, 30.6% had school difficulties, and .4% had psychosis. The follow-up studies revealed that persisting disturbances most commonly followed evacuation and consequent separation from parents. After air raids acute disturbances were noted also in children who were not evacuated, but more of these recovered. However, an immediate evacuation of the child who had been recently exposed to bombing seemed to increase the psychopathology. All of the disturbances occurred in higher frequency when the parents were apparently neurotic. Since approximately 66% of these disturbed children were noticeably "nervous" before the war, the effect of the specific events of the war were only a part of the total sickness.

71

CARLSON, E. F. Project for Gifted Children; a Psychological Evaluation, *American Journal of Orthopsychiatry*, 15:648–661 (1945)

Setting: Special class

Subjects: 25 children; age 7 years 9 months to 9 years 6 months at beginning of study; I.Q.'s, 129–159

Time span: 4 years

Meth. of obs. and test.: Case histories; projective tests; I.Q.'s; clinical observations; parents' and teachers' reports

Findings

On the basis of these studies three categories were defined: well-adjusted children of superior intelligence (7 children); children with behavior problems related to or accentuated by high intellect (6 children); neurotic children whose problems had little connection with their intellectual superiority (12 children). Of the 25 children in the special class, 20 seemed to have improved appreciably in the opinion of parents, teachers, and guidance staff. Improvement was measured by comparing the progress of these children with that of gifted children in a regular school program. The children in the special class made considerably more progress; this improvement was manifested by a reduction or modification of social and emotional problems.

Author's interpretations

The author feels that the factors which influenced this improvement were the less restrained environment, the emphasis on individual development of personal aptitudes and interests, the goal-directed activities which encouraged self-discipline, and the smallness of the group which united and helped the socially timid children through common interests. A real security accrued to the child through acceptance by this group. Because the tasks were more suitable to the intellectual abilities of these children, there was a greater stimulation for ego development in the learning process.

72

CARTER, H. D. The Development of Vocational Attitudes, *Journal of Consulting Psychology,* 5:185–191 (1940)

Setting: High school

Subjects: 75 boys and 65 girls who were part of an extensive study on adolescent growth done at the University of California

Time span: 2 years

Meth. of obs. and test.: Strong Vocational Interest Blanks

Findings

The choice of vocation appears much earlier and with much more stability than one would ordinarily imagine. Although in most cases there is no very marked deviation from the original choice, when the choice of vocation changes it usually does so in the direction of adaptive reality testing.

Author's interpretations

The combined data of scores on the Strong vocational tests together with systematic evidence from growth studies provided excellent data for intelligent vocational guidance.

73

CHARLES, D. C. Ability and Accomplishment of Persons Earlier Judged Mentally Deficient, *Genetic Psychology Monographs,* 47:3–71 (1953)

Setting: Not stated

Subjects: 151 persons judged mentally deficient

Time span: 19–20 year follow-up of patients studied by Baller in 1935

Meth. of obs. and test.: Case studies; I.Q.'s; evaluations of social status

Findings

The follow-up revealed a high death rate; 40% of the males who died met with violent, though preponderantly accidental, death. There was also a fairly large amount of delinquent behavior in this group. A large percentage had emigrated from the place of birth. Most of the group were gainfully employed and had adequate housing and living conditions; fewer were in institutions and fewer were receiving social and economic aid from agencies than originally. Of the group, 80% had married, 21% had divorced. Of the married group, 80% had had children, whose mean I.Q. was 95.

Author's interpretations

At least 65% of the cases originally described as mentally deficient would have to be reclassified in follow-up as dull rather than deficient.

74

CHILDERS, A. T. Hyper-Activity in Children Having Behavior Disorders, *American Journal of Orthopsychiatry,* 5:227–243 (1935)

Setting: Clinic

Subjects: 107 cases of hyperactivity; 500 control cases of behavior problems

Time span: Not stated

Meth. of obs. and test.: Case histories; psychiatric interviews; medical examinations; psychometric tests; questionnaires; physiological studies

Findings

General Data

There were no correlations noted between hyperactivity and sex, race, or intelligence. There was a minimal correlation found between hyperactivity and early nervous troubles, but none at all with accidents or injuries. The onset of hyperactivity was almost always gradual, usually appearing well after early childhood and at least 2 years before referral to the clinic.

Group I (30 cases studied intensively for basic findings)

Frequent traits — Of the 30 children, 76% were described as having divertability of attention, 36% were overtalkative, 40% were boastful and overconfident, 26% had a desire for attention, 80% disliked being alone and preferred association with people, 20% were cruel, and 53% showed emotional instability. In addition, most of them were very restless and had sleep disturbances and little interest in reading or individual play.

Probable environmental factors — Frequent home changes accompanied by insufficient development of inhibition seemed to be one factor. Another was faulty parental management, which permitted escape from restraint and restriction. Definite and prolonged overstimulation was a third, and insecurity due to the lack of stable relationships at home or in school.

Group II (57 cases compared with 500 random cases for physical findings)

The study revealed that a small percentage of the 57 cases had more neurological findings, but there were no correlations with endocrine disturbances. It was noted that abnormal physical findings may have added to hyperactive processes without necessarily being the cause.

Groups III and IV (10 cases in each group studied for follow-up findings)

Group III — At the time of follow-up 7 years later, 8 out of 10 children no longer showed hyperactive behavior. All of this group were referred for problems of aggressive delinquency, consisting of recurrent stealing for the boys and sex misconduct for the girls. Much sullenness and resentment of restriction was observed. From this it is assumed that though earlier overactivity in the physical sense had disappeared, it had taken the form of a lack of inhibition.

Group IV — Of the children who spent 3 to 5 years in institutions after their original referral for hyperactive behavior, 1 was completely recovered at the time of follow-up, 2 were no longer overactive, and 6 were for the most part unchanged. Serious trouble, mostly with those in supervisory capacities, usually occurred after discharge from the institutions.

Follow-ups — These follow-ups suggest that tendencies toward hyperactivity coincide with tendencies toward misconduct; both occur when restraint and restrictions are lacking and/or resented.

Author's interpretations

The author concludes that these children need a curtailment of activity, with more rest, more opportunities for gaining weight, and more regularity of management such as is offered by an institutional setting. These children fare poorly under formal discipline and need freedom, patience, and security.

75

Childers, A. T., and Hamil, B. M. Emotional Problems in Children as Related to the Duration of Breast Feeding in Infancy, *American Journal of Orthopsychiatry*, 2:134–142 (1932)

Setting: Neuropsychiatric clinic

Subjects: 469 problem children

Time span: Variable

Meth. of obs. and test.: Case histories; records of breast-feeding; parents' and teachers' reports

This is a statistical study and no individual cases are cited.

Findings

Undesirable behavior manifestations were greatest among children weaned between 1 and 6 months of age. Undesirable traits in those children who were never breast-fed had next greatest frequency. The group in which breast-feeding was prolonged more than 11 months had fewest such traits.

76

CHITTENDEN, G. E. *An Experimental Study in Measuring and Modifying Assertive Behavior in Young Children* (Society for Research in Child Development, Monograph, vol. 7, no. 1). National Research Council, Washington, D. C., 1942

Setting: Nursery

Subjects: 10 experimentals; 9 controls

Time span: Not stated

Meth. of obs. and test.: Timed observations; specially developed rating scale of assertiveness

Of 19 children who were high in domination and low in cooperation, 10 were "trained" in a series of 11 15-minute play periods putting dolls in social situations and deciding (adult and subject) the nature of the situation and the appropriate responses to it.

Findings

The trained children were significantly less dominative and more cooperative immediately after training than the controls. The increase in cooperation was no longer significant a month later. The total social activity, determined by test records and preschool observations, did not decrease although the decrease in domination was relatively greater than the increase in cooperation. This decrease in domination was not accompanied by an increase in submissiveness.

Author's interpretations

It is not possible to say that the change was due chiefly to training.

77

COLEMAN, R. W., KRIS, E., and PROVINCE, S. The Study of Variations of Early Parental Attitudes, in: *Psychoanalytic Study of the Child* (vol. VIII). International Universities Press, New York, 1953

Setting: Clinic, hospital, and home

Subjects: 4 mothers and their infants

Time span: Prepartum and first year of life

Meth. of obs. and test.: Social case work interviews; developmental examinations; intensive prenatal examinations; pediatric examinations; observations of testing through one-way mirror

The authors present four case histories and describe the earliest changes in parental attitudes.

Findings

In two cases the mother's initially close relationship with the child was negatively influenced by his growth and development. In the other two cases there was a transition from negative to positive attitudes on the mother's part as the child grew and developed.

Authors' interpretations

Parental attitudes toward the child are continuously influenced by the child's growth and development. Emphasis is laid upon the adaptive element in parent-child relationships, and the point is made that its impairment should be recognized as a factor in early diagnosis of present and/or expected difficulties.

78

Cushing, H. M. A Tentative Report of Influence of Nursery School Training on Kindergarten Adjustment as Reported by Kindergarten Teachers, *Child Development,* 5:304–314 (1934)

Setting: Kindergarten

Subjects: 33 children who attended nursery school

Time span: 1 semester

Meth. of obs. and test.: Rating scales

Findings

There were no striking differences observed between the nursery school children and non-nursery school children of similar chronological ages. There were some minimal differences in the nursery group's total adjustment to the school situation, i.e., children with nursery school experience were somewhat better in their general attitudes toward kindergarten. The mothers of the children who had been in nursery school seemed much less demanding of the kindergarten teachers.

Author's interpretations

The author feels that much more research is needed before one can speak authoritatively about the subsequent progress and adjustment of children exposed to a nursery school experience.

79

DENNIS, W. The Effect of Restricted Practice upon the Reaching, Sitting, and Standing of Two Infants, *Pedagogical Seminary and Journal of Genetic Psychology,* 47:17–32 (1935)

Setting: Author's home

Subjects: Female twins reared with complete rigidity of schedule and without toys or social responses

Time span: First 14 months of life

Meth. of obs. and test.: Developmental examinations; timed observations

Findings

Forms of motor behavior which were retarded beyond the usual upper age limit were visual motor behavior involving reaching and grasping, sitting alone, and standing with help (i.e., holding on to something). These responses were readily established once practice was instituted and were easily perfected without social encouragement or approval.

80

DENNIS, W. Infant Development under Conditions of Restricted Practice and of Minimum Social Stimulation, *Pedagogical Seminary and Journal of Genetic Psychology,* 53:149–157 (1938)

Setting: Author's home

Subjects: Female twins reared with complete rigidity of schedule and without toys or social responses

Time span: First 7 lunar months

Meth. of obs. and test.: Developmental examinations; timed observations

Findings

During most of the first year and under conditions of minimum social stimulation and restricted practice, the children yielded a development record not distinguishable from that of infants in normal environments. No general retardation appeared in one subject, while the general retardation of the other subject from the age of 10 months was referable to a birth injury.

81

DENNIS, W. Infant Development under Conditions of Restricted Practice and of Minimum Social Stimulation, *Genetic Psychology Monographs,* 23:143–191 (1941)

Setting: Author's home

Subjects: Female twins reared with complete rigidity of schedule and without toys or social responses

Time span: From 2 to 14 months of age

Meth. of obs. and test.: Developmental examinations; timed observations

Findings

In these children brought up without any initiating social responses from the experimenters who reared them, smiling and laughing were neither imitative nor the result of a social learning process. Apparently, the twins consistently made adequate social responses to the experimenters even though the latter did not return the responses; however, certain of the motor functions such as sitting and reaching seemed to be delayed. It should be noted that once the twins began to smile and laugh, the experimenters responded in turn.

Author's interpretations

The author concludes that in a child's first year most so-called social behavior is the result of a maturative rather than a learning process. He describes it as "autogenous." The question as to whether social responses might have been erased had there been no response by the experimenters has not been dealt with in this experiment. The author does feel that the social development of these children was normal at the end of the experiment.

82

Despert, J. L. A Comparative Study of Thinking in Schizophrenic Children and Children of Preschool Age, *American Journal of Psychiatry*, 97:189–213 (1940)

Setting: Nursery school and psychiatric institute

Subjects: 19 normal children (age 1 year 11 months to 5 years 3 months; average I.Q., 132); 3 psychotic patients (age 7 to 14 years)

Time span: Normal group, 1937–38; psychotic group, 1934–37

Meth. of obs. and test.: Normal group — developmental histories, psychometric tests, daily reports from home, play analysis, daily school observations, medical histories; psychotic group — psychiatric case histories, medical histories, psychometric tests, play analysis with verbatim notes on verbalizations

Findings

Group differences — There were real differences of thought and reality appreciation between the two groups. No true hallucinatory or delusional material was found in the normal group's fantasies, which could be divided into three types — denial of character of reality, evasion, and reiteration with apparent belief.

Normal child behavior — Autistic thinking is associated with a loss of contact with the external world of reality, whereas fantasies do not contribute to the child's separation from external reality. The schizophrenic's affective interests are withdrawn from reality, but the normal child retains active, affective contact with the environment, corresponding to his age level. The child shows greater emotional lability and greater susceptibility to somatic changes than the adult. Pseudo-hallucinations are dependent on the nature and intensity of the child's emotion.

Author's interpretations

It appears evident that "experiences which most closely resemble those found in the schizophrenic are dependent upon emotional factors and not upon characteristics inherent in child thinking."

83

Despert, J. L. A Method for the Study of Personality Reactions in Preschool Age Children by Means of Analysis of Their Play, *American Journal of Psychology*, 9:17–29 (1940)

Setting: Nursery school

Subjects: 15 normal children divided into 3 groups — age 2 to 3 years, 3 to 4 years, and 4 to 5 years respectively

Time span: Not stated

Meth. of obs. and test.: Developmental histories; pediatric examinations; Stanford-Binet Intelligence Scale; Merrill-Palmer Scale; behavior records of 113 ratings; observations in school and home; time-sampling study of play and play situations; selected experimental play situations; special triannual reports

Findings

In doll play, and to a lesser extent in drawing, the child dramatized and verbally expressed chiefly his affective relations to his family. The experimenter was often incorporated into the dramatization; a frequent change of role for the child or experimenter was noted. The fantasy expression was well individualized. Aggressive reactions were the most prominent in doll play, although anxiety reactions were also noted. Curiosity was expressed about body functions and sex. Excitement increased the frequency of excretion in 11 out of 15 subjects, and speech deviation occurred when excitement was coincident with aggression, anxiety, and sexual expression. Children showed individual differences in the same play situation, and varying degrees of parallelism or divergence between affective reactions in a social group and affective reactions in a play situation. This points to mechanisms of social inhibition, repression, integration, and successful adjustment. The personality investigation depends on a comparison of experimental findings with anamnestic and behavioristic data. The work of finding correlations is proceeding slowly toward this goal.

84

Despert, J. L. *Preliminary Report on Children's Reactions to the War, Including a Critical Survey of the Literature.* Pamphlet published by New York Hospital and Cornell University Medical College (Dept. of Psychiatry), New York, 1942

Setting: Nursery school

Subjects: 57 children

Time span: Variable

Meth. of obs. and test.: Questionnaires

Findings

All children who showed war anxiety had been anxious in nursery school. This anxiety was often characterized by compulsive tendencies. A child's particular reactions could almost be predicted in terms of his prewar personality. Anxious children who could discharge tension through aggression were more apt to develop antisocial patterns of behavior and less apt to display the kind of behavior which would be described as "war anxiety." The younger children did not seem to be affected, probably because they did not intellectually comprehend war, and these subjects had had no actual personal experiences with it.

85

Despert, J. L. Emotional Factors in Some Young Children's Colds, *Medical Clinics of North America,* 28:603–614 (1944)

Setting: Nursery school

Subjects: 63 children; 32 boys, 31 girls; age 2 to 5 years

Time span: 5 years

Meth. of obs. and test.: Pediatric examinations; play analysis; parents' records; school records of absence due to colds

Findings

Incidence of colds — The number of days children were absent from the nursery school varied from 1 to 35; the average number of days absent because of colds varied from 1 to approximately 12.5. The number of colds per school year for any one child varied from 1 to 16. The seasonal incidence of colds was plotted graphically and curves showed a cyclic character with peaks in winter and spring. The maximum and minimum periods of colds fell in different months every year. There is some indication that boys have colds more frequently than girls.

Emotional factor — The incidence of colds was 19.05% for 8 children from broken homes as compared with 9.46% for 8 happy, well-adjusted children. The children from broken homes were physically well-developed and well-nourished and had good health records except for the variously reported colds. Six were only children and 2 had siblings, 1 younger and 1 older. In all 8 cases the marital break had been acute, the child staying with the mother and having infrequent contact with the father. At school they often fancied "daddy's experiences," or were in some way "different from [the] group." Their play reflected anxiety and frustration with hostile attitudes toward other children and adults. The author further noted that the children from intact homes who had the highest ratio of absences due to colds also tended to show emotional stress. However, there were a few children with frequent colds who were apparently free from emotional stress, and some with definite emotional stress who were relatively free from colds.

Author's interpretations

The author feels "that in some cases internal tensions due to psychologically traumatic situations may be operating among other factors contributing to individual cold susceptibility."

86

DESPERT, J. L. Urinary Control and Enuresis, *Psychosomatic Medicine*, 6:294–307 (1944)

Setting: Nursery school

Subjects: 60 children; age 2 to 5 years

Time span: Variable

Meth. of obs. and test.: Psychiatric case histories; family histories; daily behavior records; developmental data; play sessions with psychiatrists; daily home reports

Findings

Bladder training — Training began at 12.8 months on the average and was usually achieved between 1 and 3 years; daytime training was achieved by 21.4 months and night training by 27.3 months. Thus the average child took approximately one year to be trained. Methods of training reflected the mother's directness of purpose or lack of it, her own adjustment, her feeling of security or insecurity, and her conscious and unconscious motives. In 6 cases the child's training could be directly related to the mother's own training. The children reacted to being wet in different ways, but those who were uncomfortable and objected early were trained early. This group weighed more at birth on the average, were sturdy and active, ate well, and showed early psychomotor development. On the other hand, children trained by or before 1 year of age were too well behaved generally and tended to express their hostility and hyperactivity in episodic outbursts of excitement. In comparison with the loosely organized enuretics, they were described as overorganized.

Enuretic children — For this paper a child was considered enuretic when he lacked daytime control at age 2½ and nighttime control at age 3. On this basis, 14 (10 boys, 4 girls) of the 60 children were considered enuretic; all but 1 had begun training late. These children had average birth weight, were good eaters, and gained weight. However, all sat and walked late and showed infantile mannerisms such as thumb-sucking. They were quiet babies and had difficulty in expressing outwardly their aggressive impulses. Three boys had nightmares and temper tantrums which seemed to be precipitated by their resentment and aggressive impulses. The analysis of their fantasy life bore out the conclusion that aggression was of frequent concern to these children. The author suggests that enuresis is related to urethral eroticism and sadism.

Author's interpretations

It is the author's opinion that inconsistency in the training environment was the most common cause of enuresis. She advises that training be begun not earlier than 18 months of age and that it be carried out by the person with the most consistently satisfactory emotional relationship. She suggests that the double learning process for boys (urinating while standing and defecating while sitting) may account for the preponderance of enuresis among boys. Also she notes that any change or discontinuity in the environment may cause relapses; for example, the birth of a sibling often precipitated a relapse. In discussing the data, the author points out that bladder control depends on the myelinization of the neural pathway. She discusses urinary retention also as a symptom related to interactions with the mother. Enuresis is considered in part a substitute for blocked aggressive attitudes.

87

DESPERT, J. L. Anxiety, Phobias, and Fears in Young Children, *Nervous Child,* 5:8–24 (1946)

Setting: Nursery school

Subjects: 78 preschool children; 35 anxious, 43 nonanxious

Time span: 3 years

Meth. of obs. and test.: Pediatric examinations; records of play; observations in home; play analysis; records of specific fears

This study emphasized prenatal and neonatal data but also included material on family histories and relationships, development and training, fears manifested in concrete and fantasy situations, obvious symptoms of anxiety and fear, protective mechanisms against anxiety, and finally the child's progress during the 3-year period at school. Comparison of these records with the problem behavior of the parents gave a basis for interpreting the etiology and prognosis of certain behavior disorders.

Findings

Birth factor — Of the total, 28 anxious and 17 nonanxious children had unfavorable birth conditions; 11 anxious and 2 nonanxious children came from broken homes. In both groups the mother's age at the time of the child's birth was higher than the average for the general population.

Phobias — In the preschool age group, two types of phobias could be distinguished. The first included eating and biting animals (foxes, wolves, turtles, birds); the second, which appeared later in the 2 to 5 year range, included overpowering animals and powerful people (bears, lions, giants, witches).

Author's interpretations

Unfavorable birth conditions (everything between ideal delivery and gross birth injuries) are considered one important factor in the genesis of anxiety in young children.

88

DESPERT, J. L., and PIERCE, H. O. The Relation of Emotional Adjustment to Intellectual Function, *Genetic Psychology Monographs,* 34:5–56 (1946)

Setting: Nursery school

Subjects: 39 children

Time span: 1 year

Meth. of obs. and test.: Psychiatric case histories; pediatric examinations; Stanford-Binet Intelligence Scale; play analysis; observations in school

Lengthy case summaries are presented.

Findings

When 12 children improved in I.Q., they also improved in emotional and social adjustment in the same period. And when 10 children dropped in I.Q. they showed emotional difficulties which were often attributed to the onset of sibling rivalry. Those who had little change in I.Q. had a secure, stable environment.

Authors' interpretations

The authors conclude that the total emotional balance has an important effect on I.Q. changes.

89

DUREA, M. A. A Survey of the Adjustment of School Children, *Child Development,* 10:107–114 (1939)

Setting: School

Subjects: 1,838 students

Time span: 7 months

Meth. of obs. and test.: Yepsen Adjustment Score Cards

Findings

It was found that the problem of adjustment was one of individual rather than social class differentiation; no significant differences were found between white and Negro children, or between boys and girls. In general the adjustment scores were negatively skewed. There was some evidence that the scores may have been affected by the attitude of the teacher toward the pupil.

90

ESCALONA, S. K. Feeding Disturbances in Very Young Children, *American Journal of Orthopsychiatry,* 15:76–80 (1945)
Setting: Nursery in a reformatory for women
Subjects: Nursery children
Time span: 20 months
Meth. of obs. and test.: Developmental records; general observations

The data in this article are presented as the author's observations.

Findings

While infants under 5 months of age usually made good initial adjustment to the nursery, children over 5 months had acute emotional upsets and strikingly disturbed eating habits including food refusal and projectile regurgitation. The author related these disturbances mainly to isolation from the mothers and noticed that slight and transitory emotional upsets were almost invariably reflected in the eating behavior of the children — even those who appeared to be generally well adjusted. The children responded to separation from their substitute parents as an emotional trauma; this situation was reflected in severe eating problems. Thus, it seemed that the eating situation was essentially a thermostat of the prevailing emotional atmosphere. An interesting response was observed by the author: After noting that some babies under 4 months of age abruptly reversed their preferences for tomato or orange juice, the author discovered that this reversal was coincident with a change in the personnel feeding the babies. Upon further investigation it was observed that the babies had the same dislikes as the persons newly assigned to them.

91

ESCALONA, S. K. The Predictive Value of Psychological Tests in Infancy; a Report on Clinical Findings, *American Psychologist,* 3:281 (1948)

Setting: Menninger Clinic

Subjects: 51 infants studied prior to adoption

Time span: Variable

Meth. of obs. and test.: Gesell Developmental Schedule; Cattell Infant Intelligence Scale; Revised Stanford-Binet Scale; general observations

The correlation between the estimates of intelligence obtained in early infancy and those taken during preschool years is "better than published reports indicate." The data suggest that certain aspects of mental functioning can be more accurately predicted than others. The same formal test results may be interpreted differently in the light of concomitant clinical observations, thus increasing the predictive value of the tests. It is felt that the greater stability of some aspects of mental functioning as compared with others can be explained if a "predisposition" toward certain modes of psychological experience is assumed to be present at birth.

92

EZEKIEL, L. F. Changes in Egocentricity of Nursery School Children, *Child Development,* 2:74–75 (1931)

Setting: Nursery school

Subjects: Number not stated

Time span: 3 months

Meth. of obs. and test.: Quantitative records of play and social behavior

Findings

Expression of egocentric behavior — Egocentric behavior was expressed by the type of play (alone or with others), the methods of retaining toys another child wanted, and the methods of gaining attention from other children and from adults.

Influence of social situation on behavior — Those children who were dominantly egocentric on entering a new social situation made few significant changes during the first three months, but the unaggressive children became more egocentric and aggressive in their play.

Correlation between age and egocentricity — A positive correlation was seen between the age of the child when he entered a new social situation and the type of egocentricity.

93

FITE, M. D. Aggressive Behavior in Young Children and Children's Attitudes toward Aggression, *Genetic Psychology Monographs*, 22:151–319 (1940)

Setting: Child development institute

Subjects: 6 children; only 4 are used in the analysis of data

Time span: 1¼ years

Meth. of obs. and test.: Timed observations; teacher's records; interviews; 2 experimental situations

The experimental situations were devised to provide data on the child's attitude toward aggression. The first, which was to test the child's attitude toward aggression in the form of hitting, consisted of a set of 12 pictures in 2 series, one given at each sitting and administered much like the T.A.T. The second experiment, which was devised to evoke the child's aggression, utilized blocks and dolls named after children in the nursery school group. The examiners then questioned the child with specific reference to his aggression toward the dolls. The children were exposed to these experimental situations three times. The data enabled the tester to compare the child's behavior in free play with his verbal attitude toward specific situations in a set of pictures and also with his behavior in playing with dolls and blocks. This paper also includes an excellent survey of the literature on the subject.

Findings

Results of experimental situation — In the devised situation

2 of the 6 children tended to swing from opposite extremes to the mean during the testing period. A third maintained his balanced behavior the entire period. Of the 3 younger children, one seemed to enjoy aggression and this child's closest friend seemed to base his moral attitude entirely on the other's aggressive behavior. At the beginning of the test period the third child in this group seemed completely confused about the rights and wrongs of hitting other children, but by the end of the experiment she indicated, verbally at least, that she did not like to be hit.

Comparison of experimental data with free play — Of the older children, one who had verbalized early and acted out a strict morality and had eventually approached a mean in the experimental situation changed from being quite aggressive to being less aggressive in free play over the same period of time. Thus as his verbal attitude became freer, he became less aggressive in free play. In contrast, a second child who developed a strict morality during the experimental time became increasingly aggressive in the free play period. Similar but less clear-cut trends were noted in the younger children.

Types of aggression — The author noted that there were differences in the types of aggression displayed over a period of time. Thus the children who had become less aggressive showed their aggression in a more determined and persistent way. The author concluded that there was no relationship between what children said about aggression and what they did. Thus these children in their attitudes about anger displayed the dual standards described by Anna Freud.

Other factors influencing behavior — The child's social role, social intimacy, and maturity seemed to affect his aggressive behavior, which was in part a manifestation of his overall personality. Thus, one child's aggressiveness was traced to his feelings of insecurity.

Teacher's influence on behavior — The author described the nursery school teacher's attempts to modify aggressive behavior. In most cases she tried to make the children talk about aggressive tendencies instead of acting them out, but the children tended to talk in a much more destructive way than the teacher had anticipated. The teacher had the most beneficial effect on aggression when she was able to help the children manipulate difficult situations rather than when she found it necessary to restrict or reprove children. Rules imposed by adults were employed by the children as aggressive instruments. On the other hand, they used social patterns to replace overt aggression.

Influence of parents' attitudes (based on interviews with parents) — The child reflected his parents' rules and attitudes in discussions with the interviewers but not in play with dolls. When it was suggested to the child that he ignore his parents' rules about aggressive behavior, strong emotional responses were encountered.

Consistency of aggressive attitude — The author noted no consistent developmental attitude about aggression. However, in the nursery school environment children seemed to promote more aggression in defense of their rights and independence. Contacts with other children developed the following: a tendency to regard aggression more leniently than adults, group and friendship mores which tended to supplant adult mores but which remained somewhat dependent on them and tended to modulate aggressive behavior, and group status which had a marked effect on aggressive behavior.

94

FOSTER, J. C., and ANDERSON, J. E. *The Young Child and His Parents; a Study of One Hundred Cases* (Institute of Child Welfare, Monograph Series, no. 1). University of Minnesota, Minneapolis, 1930

Setting: Institute

Subjects: 100 children; age 2 to 6 years

Time span: 1 to 4 years

Meth. of obs. and test.: Interviews; case histories; I.Q. tests

The 100 case histories have been grouped to bring out salient features of this study of normal children in the home situation. Summaries of the findings are presented in tables.

Findings

Proportion of various types of problems — By far the greatest number of problems related to emotional adjustment. Nervous habits, feeding difficulties, and conflicts with authority were found in one-fourth of the cases. Overdependence, overimaginative behavior, handling genitals, improper language, and school difficulties were found in less than one-tenth of the cases. Boys had a mean of 3.1 problems while girls had a mean of 2.8.

Development of problems — Tantrums were at their maximum at age 2, fears and sleeping problems at age 3, feeding problems at ages 3 and 4, playmate trouble at age 4, and fatigue at age 5. Authority problems increased between the ages of 2 and 4. Half the children aged 4, 5, and 6 had emotional problems. Analysis of the cases after a lapse of 1 to 4 years showed a marked tendency for problems to decrease as the child grew older and moved into the social environment of school and community. On the basis of home rating, problems decreased or disappeared in "good" homes and few new ones arose. In families rated "poor" a fair percentage of old problems persisted and a number of new ones appeared.

95

FOSTER, S. A Study of the Personality Make-up and Social Setting of Fifty Jealous Children, *Mental Hygiene,* 11:53–77 (1927)

Setting: Habit clinic

Subjects: 50 jealous children and 100 nonjealous controls

Time span: Not stated

Meth. of obs. and test.: Case histories

Tables, case histories, and a discussion of literature on the subject are presented in addition to the findings of the study.

Findings

Age and sex distribution — Of the jealous children, 2 out of 3 were girls, while the sex distribution was equal in the control group and among clinic patients in general. The largest number of jealous children were 3- and 4-year-olds who were beginning to show signs of individualism which was frequently associated with contrariness. In 27 out of 50 cases, the jealous child was the eldest in the family.

Home environment — Friction was present and good training was lacking in the homes of both groups. One-third of the children in each group had at least one neurotic parent. Of 50 jealous children, 28 were subjected to physical punishment while only 23 of the 100 nonjealous children were so treated. The author felt that jealousy was aroused in children partly because they resented punishment and compared the amount of punishment they received with what others received. In the jealous group, 8 children (all girls) showed marked attachment for the mother, while 6 (2 boys and 4 girls) showed attachment for the father. The determining factor seemed to be the amount of attention the child received from the individual parent. Teasing from parents, siblings, and others was much more evident in the jealous group.

Personality traits and habits — Confidence, selfishness, pugnacity, and capriciousness about food were displayed by jealous children. They also suffered from emotional tension and sleep disturbances. Enuresis was present in one-half of the jealous group but in only one-third of the control group. More nail-biting and thumb-sucking but less masturbation were found in the jealous group. Both groups were limited in their play, but the jealous children were especially limited in their social contacts in play. There were no correlations with general intelligence.

Author's interpretations

The author summarizes by presenting a profile of the personality and home situation of a jealous child. This child, though often active and domineering in his own environment, plays poorly with others due in part to inadequate, unintelligent training and play opportunities. He is likely to be thrust aside at the birth of a sibling and he may show overt signs of jealousy.

96

FRENKEL-BRUNSWIK, E. Motivation and Behavior, *Genetic Psychology Monographs,* 26:121–264 (1942)

Setting: Not stated
Subjects: 150 students
Time span: Grade 5 to high school graduation
Meth. of obs. and test.: "Intuitive" ratings made by judges in terms of categories devised by H. A. Murray

The author tries to correlate drive ratings with one another, with manifest behavior, and with self-reports (Adjustment Inventory). She also evaluates the relationship of rated drive patterns to environmental data.

Findings

Different types of behavior were often related to one drive as alternate but unrelated manifestations of the same set of dynamic factors. For example, they described two clusters of behavior — "overt social activity" and "emotional maladjustment." Adolescents who rated high on the aggressive drive cluster (social ties, recognition, control, or escape) might be either emotionally maladjusted, successful in overt social activity, or both. Other patterns were found when ratings of manifest traits (emphasizing especially consistent attitudes, habits, actual adjustment) were correlated with rated drives. Certain manifest features, especially those concerned with adjustment, security, and maturity, originated in a variety of underlying motivational conditions. Relations between motivational patterns and overt behavior were ambiguous, but the indeterminateness of the relationship was somewhat reduced if the *total* patterns were related to behavior patterns. Drive ratings showed a good correlation with independent data such as self-reports.

Author's interpretations

The author comments that the study justifies the concept of motivation and the status of drive ratings used here. Correlations indicate that ratings genuinely revealed dynamic states of the subject and not notions of raters. The subjective factor, however, is analyzed and discussed and it was found that these factors tend to cancel themselves out. The drive ratings are therefore considered to have predictive value.

97

FREUD, A., and BURLINGHAM, D. T. *Infants without Families; the Case for and against Residential Nurseries.* International Universities Press, New York, 1944

Setting: Residential nursery

Subjects: Children from birth to age 10; number not stated

Time span: 10 years

Meth. of obs. and test.: Psychiatric interviews; general observations

Authors' interpretations

Children do not function as well in residential nurseries as in their families. Artificial families are a partial solution to the problems, but there is no substitute for the father. Children tend to fill the void with fantasy productions. Health, skills, and social responses are generally unaffected but children are limited in the realm of their emotional life and character development. The authors emphasize the point that early contact with other children increases aggressive responses and contributes to a surprising range of emotional reactions.

98

FRIED, R., and MAYER, M. F. Socio-Emotional Factors Accounting for Growth Failure in Children Living in an Institution, *Journal of Pediatrics,* 33:444–456 (1948)

Setting: Children's home

Subjects: Not stated

Time span: Not stated

Meth. of obs. and test.: Wetzel Grid; medical histories; general observations

Authors' interpretations

Most children with growth failure showed socio-emotional maladjustment at one point or another during their residence in the institution. In these cases maladjustment and growth failure began and ended together. Dissociated failure did occur but less frequently; the two variables were sometimes sequential. The authors use four case histories to illustrate the relationship between emotional and growth maladjustment. They believe that socio-emotional adjustment is much more critical in growth failure and ill health than was previously realized.

99

FRIES, M. E. The Study of the Emotional Development of Children, *Medical Woman's Journal,* 43:199–202 (1936)

Setting: New York Infirmary

Subjects: Newborn infants

Time span: 8 years

Meth. of obs. and test.: Psychiatric interviews; prenatal home visits; perinatal observations; pediatric examinations; motion pictures

The psychological impact of the environment on the prenatal course is illustrated by case histories.

Author's interpretations

The psychological development of children can be traced from the beginning of pregnancy in terms of the mother's reaction to it. These events may be predictive in certain instances of later mother-child relations.

100

FRIES, M. E. Factors in Character Development, Neuroses, Psychoses, and Delinquency; a Study of Pregnancy, Lying-in Period and Early Childhood, *American Journal of Orthopsychiatry,* 7:142–181 (1937)

Setting: Hospital, home, clinic, and nursery

Subjects: 47 infants selected at 6 weeks of age; an additional group chosen prenatally

Time span: Not stated

Meth. of obs. and test.: Pediatric examinations; psychiatric interviews; home visits; study of confinement; motion pictures; prenatal observations; well-baby clinic records

A case history is presented which contains a sample of the data. [The father described in this history appears to resemble closely a description by Phyllis Greenacre of a personality type having the unconscious desire to give birth to a son.]

Findings

Some important individual factors affecting personality development were described by the author. First was the parent-child relationship. Also there seemed to be fundamental differences between children even in the first 10 days of life in such areas as the amount of activity, sensitivity, reaction to frustration, sleep behavior, etc. These seemed modifiable but not extinguishable and persisted until 6 months of age. Too, the degree of maturity at birth seemed to affect the rate of development. Maturity was judged by a calculation of the gestation period, an evaluation of general behavior, physical examinations and measurements, x-rays of the carpal bone, and neurological examinations. Related to maturity was the extent of myelination present at birth, for which the Moro Test seemed the best indicator. Lastly, socio-economic status had an important effect on the child's development because it seemed to be closely related to child-rearing practices.

Author's interpretations

The author presents the following observations: Almost all the mothers in the study seemed to have a poor understanding of prenatal factors and influences. Psychotherapy for these parents was helpful early in the child's life, particularly with respect to feeding difficulties, and was more effective when directed toward the father as well as the mother. Parents tended to be more cooperative early in the child's life than later on. The overall approach provided by this study's facilities seemed to decrease obstetrical and pediatric care requirements. In addition, it seemed to provide new insights into the etiology and development of character and to facilitate earlier detection of maladjustment, thus providing, apparently, an easier and more successful therapy. The author feels that the periodic predictions provided an opportunity for evaluative summaries and scientific consideration of their past and present studies. These predictions indicated patterns and trends to be observed, often demonstrated approaching difficulties which could be avoided, and ultimately seemed to provide a basis for prophylactic measures.

101

FRIES, M. E. The Value of a Play Group in a Child-Development Study, *Mental Hygiene,* 21:106–116 (1937)

Setting: Pediatric clinic

Subjects: 3 children

Time span: Not stated

Meth. of obs. and test.: Pediatric examinations; psychiatric interviews; psychological examinations; observations during play; home visits

The reports of the play of three children are given to illustrate the wealth of material to be derived from the observation of play. The mothers were given group therapy concurrently.

Author's interpretations

The value of a play group is that it gives children the incentive to attend the clinic, reveals adjustment to peers and reactions to deprivation and indulgence, provides insight into mothers' reports and family relations, points to steps for therapy, and provides therapy by group activity.

102

FRIES, M. E. The Importance of Continuous Collaboration of All Agencies in Dynamic Handling of the Child, *Nervous Child,* 3:258–267 (1944)

Setting: Hospital, home, clinic, and nursery

Subject: 1 child

Time span: 11 years

Meth. of obs. and test.: A case record compiled on specially devised charts

In this paper the investigator presents a longitudinal study of one child's developmental pattern. The article illustrates the significant data and increased understanding derived from this multifaceted approach to child care and study. In this case the background material is used to explain the child's behavior during the stress of a false air raid alarm.

103

Fries, M. E. Psychosomatic Relationships between Mother and Infant, *Psychosomatic Medicine,* 6:159–162 (1944)

Setting: Hospital and clinic

Subjects: Not stated

Time span: 5 years

Meth. of obs. and test.: Ratings on activity patterns; motion pictures

The author examines the behavior of infants in terms of gross classifications of activity. She relates the amount of activity to specific effects on the mother. Two tests are used as aids in formulating the activity patterns.

Findings

In the first test the infant's startle response was noted by direct observation and movies. Active children showed the greater response. In the second test (i.e., oral test with nipple), an evaluation was made of the infant's response to the presentation, removal, and restoration of an object of gratification. Later on other tests were evolved to measure this response. The quiet child took the nipple slowly and when it was removed continued to suck and finally fell asleep. The active child either refused the nipple or took it eagerly; when it was removed he might show a startle response.

Author's interpretations

The above patterns of response, the author argues, are prototypes of personality development. The quiet child calls upon the environment for help and should be encouraged to act; he will also take longer to arouse to nurse and so requires patience. She further states that such a baby can tolerate and should have more stimulation. The active child needs more comforting and maternal protection against excessive stimulation, more watching, and greater tolerance in order to prevent self-harm or harm to his environment. An observation period of 24 hours is required to obtain a true picture of activity patterns, which have to be isolated from effects due to myelinization. However, the amount of maturity at birth in itself has impact on the child's environment. Studies five years later showed that normal children had only slightly modified their activity patterns; children of emotionally maladjusted parents tended to move further from the median in activity.

104

FRIES, M. E., BROKAW, K., and MURRAY, V. F. The Formation of Character as Observed in the Well Baby Clinic, *American Journal of Diseases of Children,* 49:28–42 (1935)

Setting: Well-baby clinic

Subjects: 19 male and 26 female infants

Time span: 6 years

Meth. of obs. and test.: Pediatric examinations; mothers' records of sleep, feeding, and toilet training; home visits; interviews with mothers about their daily regime; observations of infants and mothers

The focus of the paper is on methods of habit training in relation to the mother's behavior, the age at which training is begun, and the condition of the infant before training is begun.

Findings

This group was given the following pediatric advice: feed the infant every 3 hours, begin toilet training when baby can sit, and wean to cup at 10 to 12 months. Then the actual transitions of habit patterns were recorded — the average breast weaning occurred at 4½ months, the change from three-hourly to four-hourly feedings and the omission of the 2:00 a.m. feeding was accomplished between the ages of 3 and 4 months, and weaning from the bottle at 10½ months. Of this group 6 children had marked behavior problems before 18 months of age; the deviation was usually noted in several areas and often physical and psychological deviations occurred together.

Authors' interpretations

It appears that deviations were more common in children with neurotic mothers than in those with non-neurotic mothers. It is the authors' conclusion that ignorance of method is less dangerous to the process of child rearing than instability in mothers.

105

FRIES, M. E., with LEWI, B. Interrelated Factors in Development; a Study of Pregnancy, Labor, Delivery, Lying-in Period and Childhood, *American Journal of Orthopsychiatry*, 8:726–752 (1938)

Setting: New York Infirmary

Subjects: Not stated

Time span: Up to 3 years

Meth. of obs. and test.: Detailed, protracted observations at birth; daily postnatal moving pictures; psychiatric interviews with all members of the family; standardized Moro and oral frustration tests; pediatric examinations

On the basis of detailed observations at birth the children were divided into activity types — quiet, moderately active, and active. Predictions were made on the basis of the activity type, what was known about the child's environment, and the reaction to the standardized tests. The authors feel that evaluation of the child and his environment permits a synthesized diagnosis which is of great value in mental hygiene. They give a theoretical combination of child and environment which might be used to predict later development. The combination includes a child who deviates excessively from the average in physical condition and maturity of the nervous system, is very quiet, becomes passive and withdrawn when deprived or thwarted, develops early feeding difficulties, and shows extreme deviant behavior throughout childhood and adolescence. The environment is extremely unsatisfactory — one, for example, in which the child is subjected to repeated deprivations due to rejection, domination, or both; or one in which the child identifies with a parent having schizoid traits.

106

Gesell, A. L. The Influence of Puberty Praecox upon Mental Growth, *Genetic Psychology Monographs,* 1:511–539 (1926)

Setting: Yale Clinic of Child Development

Subjects: 2 girls; one who began menarche at age 3 years 11 months (studied until age 7); the other who began menarche at age 8 years 3 months (studied from age 6 to 10)

Time span: Not stated

Meth. of obs. and test.: Medical histories; anthropometric data; "Draw-a-Man" test; vocabulary; play interests; school reports, etc.

Findings

In these two subjects precocious puberty was not accompanied by a coordinate spurt of mental growth. One of the subjects was a mental defective who was in the longitudinal study when menarche occurred. Menarche had no effect on her mental development, but she showed an intense shame reaction which was thought to be instinctive. The author believes that precocious puberty "alters psychic patterns and introduces effective alterations in the attitudes and in the temperamental susceptibilities." The change is a real disturbance, but it "concerns personality rather than intellect."

Author's interpretations

The author feels that puberty praecox had an overall dislocating effect on the "total growth complex" particularly in the emotional sphere.

107

Gesell, A. L. *Infancy and Human Growth.* Macmillan, New York, 1928

Setting: Yale Clinic of Child Development

Subjects: 90 infants; numerous case studies

Time span: Mainly the first 30 months of life, but some data beyond that point

Meth. of obs. and test.: Developmental examinations; observations through one-way mirror; cinemanalysis; studies of drawings; medical data

This text represents the formulation of data resulting in the Gesell Developmental Schedules. The methods of study employed by the Gesell group are described and examples of typical and atypical growth patterns are cited. Monthly increments in normal development are presented as the basis for a schedule to be used for normative studies. It is mainly concerned with developmental sequence and presents basic data in this area.

108

GESELL, A. L. The Stability of Mental-Growth Careers (chap. VIII), in: *Thirty-ninth Yearbook: Intelligence; Its Nature and Nurture*, Part II (National Society for the Study of Education). Public School Publishing Co., Bloomington, Ill., 1940

Setting: Yale Clinic of Child Development

Subjects: 33 children

Time span: 10 years

Meth. of obs. and test.: Gesell Developmental Schedule; Stanford-Binet Intelligence Scale

The author presents nine illustrative cases all of which exemplify consistent mental-growth careers. He also sets forth a theoretical concept of intrinsic developmental reserves — i.e., errors or setbacks of development stimulate biochemical and somatic structures (reserve factors) to regulate self-correction.

Author's interpretations

Each child proved to have a distinctive growth pattern; a marked alteration occurred in only one case (from low to high average). In most cases the trend was toward improvement. Normal growth potentialities were almost clinically perceptible in the first two years; rarely were signs of normality (emotional and physical factors) delayed as long as three years. Temporary irregularities of development were usually present in preschool years and almost every case of primary feeble-mindedness could be diagnosed in the first year. Mental-growth trends were most consistent when a balance existed between endogenous and exogenous factors.

109

Gesell, A. L., Amatruda, C. S., Castner, B. M., and Thompson, H. *Biographies of Child Development; the Mental Growth Careers of Eighty-Four Infants and Children; a Ten-Year Study.* Hoeber, New York, 1939

Setting: Yale Clinic of Child Development
Subjects: 84 infants
Time span: Variable, but in most cases more than 10 years
Meth. of obs. and test.: Gesell Developmental Schedule; case material

This text presents detailed studies of 84 exceptional children observed over a period of years by the staff of the Yale Clinic of Child Development and then selected for the study on the basis of their instructive value. Part I deals with 30 cases reported on 10 years previously. These reports are reviewed and compared in the light of follow-up studies and comments are made on the possibilities and limitations of developmental diagnosis. Part II comprises 54 individual studies of mental growth including some typical and atypical forms and special social and guidance problems (twin relationships, irregular mental development, superior I.Q., foster care, language and reading difficulties, physical handicaps).

Authors' interpretations

Differences in growth are due to original capacity, rate or tempo, and patterns of developmental organization. Infants differ significantly at birth and do not indicate any standard pattern of growth. Environmental factors support, inflect, and modify but do not generate the progress of development. Behavior in the first year is consistently related to behavior in the fifth year.

110

Gesell, A. L., and Ames, L. B. Early Evidence of Individuality in the Human Infant, *Scientific Monthly,* 45:217–225 (1937)

Setting: Yale Clinic of Child Development
Subjects: 5 infants
Time span: 5 years
Meth. of obs. and test.: Motion pictures

From the cinema records of the first year of life, a monthly summary of behavior was made for each situation. Then the following list of behavior traits was made: (1) energy output, (2) motor demeanor, (3) self-dependence, (4) social responsiveness, (5) family attachment, (6) communicativeness, (7) adaptivity, (8) exploitation of environment, (9) humor sense, (10) emotional maladjustment, (11) emotional expressiveness, (12) reaction to success, (13) reaction to restriction, (14) readiness of smiling, (15) readiness of crying. The children were ranked according to each trait; at 5 years of age this was repeated.

Authors' interpretations

Internal consistency, which seems to depend on "biological characteristicness," is present to a significant degree in a child's behavior at ages 1 and 5. Since physical and cultural environmental features influence basic constitutional features, early individuality can be seen. Results indicate a high degree of latent predictability in traits seen in the first year and show that fundamental individual differences do not increase markedly with age. For 3 children, correct predictions were made prior to 16 weeks of age on traits 1, 2, 3, 11, and 14.

111

GESELL, A. L., and AMES, L. B. The Infant's Reaction to His Mirror Image, *Pedagogical Seminary and Journal of Genetic Psychology,* 70:141–154 (1947)

Setting: Yale Clinic of Child Development
Subjects: 1 child
Time span: From 16 to 60 weeks of age
Meth. of obs. and test.: Motion pictures

Findings

The longitudinal survey disclosed significant ontogenetic trends. At 16 weeks the child indicated awareness of his mirror image by the expression in his eyes; later he showed his reactions by movements of his arms, hands, feet, fingers, toes, and tongue. His initial concern with the reflection of his face was expanded to include his total image and surroundings. Perception of depth and distance and social awareness increased with age. Manipulations and explorations progressed from relatively gross to more finely elaborated coordinations. Though fluctuations occurred in the progressive expansion of the child's responses, there were consistent trends which reflected underlying developmental factors.

112

GESELL, A. L., *et al.* *The First Five Years of Life; a Guide to the Study of the Pre-School Child.* Harper, New York, 1940

Setting: Home, school, and research institute

Subjects: Not stated

Time span: 5 years

Meth. of obs. and test.: Not specifically listed by the author but apparently used were developmental examinations, interviews with mothers, naturalistic observations of play, observations through one-way mirror, cinemanalysis, developmental histories, family histories

This text presents a broad approach to a study of child development. It is divided into three parts and includes examination records and arrangements. Part I is written by Gesell and consists mainly of two outlines, one of the child's first year and the other of his next four years. The child is described at different stages in terms of four categories — motor, adaptive, language, and personal-social. Part II is a concise elaboration of the data presented in Part I. Age gradations for such behavior as eating, sleeping, and elimination are presented in Chapters 9 and 12. Part III is concerned mostly with the developmental examination. Important aspects of the personal-social category discussed in Part I are presented here.

Findings

Development in first year — At the age of 4 weeks, the child has a brief, intent social regard and responds to competent handling. At 16 weeks he recognizes his mother and smiles. Pleasure in motor exploration and, in a sense, self-sufficiency appear at 28 weeks. By 40 weeks he settles into routines, enjoys having people around, and can respond to demonstration.

Development from age 1 to 5 — By the time the child is 1 year of age, he repeats performances, begins to understand others' emotions, and has unmistakable, identifiable emotions himself. He distinguishes "mine," is resistant to sudden changes, and is defiant by 18 months. There is also an increase in imitative behavior at this point. A definite sense of self, parallel play, jokes, and feelings of guilt appear in the child 6 months later. By 3 years of age he can sacrifice immediate satisfaction for future reward. He also responds to suggestions and talks to himself; his social life broadens. The child is characteristically assertive, independent, talkative, and phobic at 4 years of age at which point he begins to care for and toilet himself. He prefers group activity, and by the time he is 5 years old he plays in larger groups. At this age, 5 years, he is independent, "adultish," and dependable, and has developed a concept of shame.

113

GESELL, A. L., and ILG, F. L. *The Child from Five to Ten.* Harper, New York, 1946

Setting: Home, school, and research institute

Subjects: Approximately 64 children

Time span: 10 years for some; 5 years for others

Meth. of obs. and test.: Developmental examinations; Stanford-Binet Intelligence Scale; performance tests including Arthur Point Scale of Performance Test; reading readiness tests including Monroe Reading Aptitude Test; visual skills; naturalistic observations of play; open-ended interviews with parents; observations through one-way mirror; family histories

The text represents a continuation of the study on the children described in *The First Five Years of Life* and the findings are presented in the form of growth gradients. The book is divided into three sections. Part I is entitled "Growth" and is a general discussion of growth and development with specific examples of gradients — e.g., acquisitive behavior is spaced from age 5 to 15 years. Part II is entitled "The Growing Child." The first chapter recapitulates data on the first four years of life; the second chapter deals with the 5-year-old and makes explicit the concept that "underlying pervasive traits constitute his 5-year-oldness." This concept is developed, indicating that there are certain maturity traits which distinguish the 5-year-old from the 4- or 6-year-old. These maturity traits are then outlined in the ten categories of motor, hygiene, emotional expression, fears and dreams, self and sex, interpersonal relations, play and pastimes, school life, ethical sense, and philosophical outlook. Part III deals with the "Total Growth Complex." The ten categories of maturity traits are considered separately in terms of age gradients. Pertinent categories from Part II are summarized here for each age group.

Findings

Age 5 years — The child is more attached than before to the mother and siblings; he has few fears but does become frightened at the possibility of losing his mother. He is poised, helpful, persistent, and somewhat shy; will hold his own ground in an argument; is more confident with people; and is a good player. His dreams are concerned with biting animals, his sexual interest with reproduction, and his play increasingly with dolls. The sense of good and bad is purely pragmatic to him.

Age 6 years — The child is characteristically egocentric and possessive at this period, and shows increased tension with rigidity blending into ambivalence. Temper tantrums and sex tend to be exhibitionistic, and sex play (the child may exhibit or undress other children) and sex interest (how the baby comes out) increase. Interpersonal relations are difficult at this time; he clashes with his parents, but tends to develop a good relationship with his father. The child fears large animals and insects and begins to be concerned with death. His fears are attributed to his having to face stimuli which he cannot comprehend or organize. He dreams more about fire and less about biting animals and begins to have nice dreams.

Age 7 years — At this age the child has more control, is concerned with self, has empathy for characters in stories, tends to withdraw from conflicts, has obsessive play interests, and becomes modest and sensitive about his body. There is less sex play but intense sexual curiosity, and a few boy-girl pairs may develop. He fights with peers, yet can be a helpful companion to adults. His standards for himself are high and his dreams about himself involve floating, flying, etc. and changing sex. He also begins to think through and rationalize his fears.

Age 8 years — Now the child becomes less sensitive and less withdrawn. He likes to argue and tends to dramatize and be impatient; although his fears are under control, he may become a worrier. His standards are based on other people's wishes rather than his own. There is a growing social concern; he is better behaved away from home and begins to develop special friends. Sexual curiosity turns to how a baby is made and girls may ask their mothers outright.

Age 9 years — He is independent and dependable, and is loyal and devoted to his friends. He is a planner, is concerned with his reputation, is easily disciplined, and responds to suggestions. The 9-year-old has few fears but frequent nightmares in which murder plays a prominent role. His awareness of sexual differences and his curiosity about genitals of the child's own sex are apparent.

Age 10 years — By this age the child is an adult in the making. He is relaxed, casual, and alert. He enjoys secrets, esteems his gang and is sensitive to its values. Gangs become unisexual and there is evidence of beginnings of adolescent concerns.

114

Gesell, A. L., Ilg, F. L., and Ames, L. B. *Youth; the Years from Ten to Sixteen.* Harper, New York, 1956

Setting: Research institute

Subjects: 115 children

Time span: 6 years

Meth. of obs. and test.: Developmental examinations; Stanford-Binet Intelligence Scale; performance tests including Arthur Point Scale of Performance Test; reading readiness tests including Monroe Reading Aptitude Test; visual skills; naturalistic observations of play; open-ended interviews with parents; observations through one-way mirror; family histories

The text is divided into three parts. Part I outlines the study and defines the concepts. Part II presents maturity profiles and traits for each age group. Part III describes growth gradients, emphasizing the transitions in each of the ten categories considered as a group at each age level in Part II. The findings related to personal and social development for the age levels in Part II are discussed here.

Findings

Age 10 years — Fears are at a low ebb, but the child may be a worrier and can become quite angry. Ambition is submerged by a desire to be part of a group. The child is poised and active, enjoys family life, and loves friends of the same sex. He has a strict moral code and sense of self-righteousness, and is concerned with fairness. Girls begin to show adolescent changes and concern with breast development; they begin to confide in their mothers, while boys turn to their fathers.

Age 11 years — There is an inner seething reflected in incessant body activity and expenditure of energy. The 11-year-old tends to be less poised and more spontaneous and immediate in his emotional expression. He is less strictly ethical and wants to set his own codes of behavior. His fears, especially of being alone, are greater, and he is apt to be quite sensitive, competitive, and rebellious. Sibling rivalry is intense, personal ambitions develop, and some cheating and stealing is evident at this time. There is a need for parents. The child is choosey about friends and boy-girl relationships begin. Girls begin to show marked variations in adolescent changes, whereas boys are relatively similar. Boys are interested in animal sexuality and have frequent erections; masturbation is known to half of them.

Age 12 years — At this age there is a calming down and the 12-year-old is able to organize his energy. There is a rapid spurt in physical adolescent changes, especially for girls who now experience menarche and begin to accept and flaunt their physical maturity; they turn to their mothers as confidantes. There is a realistic concern in practical sexual matters and heterosexual interests are present. Ethics are subject to thoughtful consideration by the child and death is of increasing concern. There is less cheating and stealing. The child develops self-reliance and self-competence. Interpersonal relations are more pleasant.

Age 13 years — There is an inward concentration of energy with more isolation and withdrawal from the family. The child turns inward in his emotions and is often sad. Ethics are now thought of as a separate body of knowledge, and right and wrong are approached with ease. The 13-year-old is not fearful, but has fearful thoughts and may be somewhat claustrophobic. He is self-critical, may be sensitive, and begins to be embarrassed by his parents. Boys have gangs, whereas girls have individual friends; there is somewhat less open interest in the opposite sex. Boys undergo a rapid spurt in physical adolescence and half of them have had ejaculations.

Age 14 years — There is more movement outward at this age with an easier give and take. Expansiveness and comprehensiveness are characteristic as is the concern over being a member of a group. Happy moods are common and are interrupted by annoyance rather than sadness. The child shows an awareness of the relative quality of ethics. His ethical sense broadens to include world problems. He has fewer fears but there seems to be a hardening of pet fears; worries are most often concerned with the body. Heterosexual feelings are now often accompanied by physical sensations and most boys have developed a pattern of sexual activity involving masturbation.

Age 15 years — The 15-year-old is somewhat preoccupied and sophomoric; tension is high. Emotions are changeable with much somberness; less competitiveness and eagerness; and insistence on freedom, independence, and privacy. Often he doesn't get along with his parents and in general family ties are loosening. The child insists on the right to his own ethical decisions and has realistic concerns about his future. Girls develop gangs, but boys have larger and more consistent gangs. There is little definite pairing off, the interaction being mostly between groups of boys and girls.

Age 16 years — The child is still sensitive but keeps to himself. There is incentive and ambition, and a remoteness from the family but now with fewer hard feelings. Ethical conclusions come easily and the child deals with them accordingly. Girls accept their sexual functions; boys have increasing difficulty controlling their sexual desires and are more sensitive to social interaction. Otherwise, emotional problems are usually in hand.

115

Gesell, A. L., and Thompson, H. Twins T and C from Infancy to Adolescence; a Biogenetic Study of Individual Differences by Method of Co-Twin Control, *Genetic Psychology Monographs,* 24:3–121 (1941)

Setting: Yale Clinic of Child Development

Subjects: Female monozygous twins T and C

Time span: Not stated

Meth. of obs. and test.: Special training for motor skills; developmental studies; cinemanalysis; analysis of drawings, language, and writing; special experimental observations; Rorschach Test; anthropometric measurements; records of habit training; posture studies; Cunningham Test; Stanford-Binet Intelligence Scale; observations of play; teacher's reports; mother's reports; open-ended interviews with mother and children; Kent-Rosanoff Free Association Test; Rogers Test of Personality Adjustment

Findings

Comparison of twins — In physical characteristics there were only minimal differences apparent. Twin T was more alert from birth in posture (motor behavior) and maintained this alertness; she was also slightly superior to twin C in intellectual ability. Twin C was more talkative and continued to be so despite special language training for twin T. Twin C was also conspicuously more sociable and was the dominant twin.

Follow-up — The likenesses and disparities observed during infancy proved durable. In spite of hundreds of hours of special training, twin T did not improve with respect to behavior, mental stature, or individuality. However, at follow-up she was more rigid, systematic, controlled, and constrained. She used straight and angular lines and gave fewer movement responses on the Rorschach Test.

Authors' interpretations

It is difficult to explain the differences between the twins on a purely psychogenic basis. The remarkable resemblance is attributed to "accurate halving of original genetic substance." The authors believe that the organism largely "determines" the environment. For example, they feel that it was because of her more sociable nature that twin C captured her mother's greater love rather than the reverse.

116

GOLDFARB, W. Infant Rearing and Problem Behavior, *American Journal of Orthopsychiatry,* 13:249–265 (1943)

Setting: Not stated

Subjects: 40 children raised in institutions for the first 3 years of life; 40 children raised in foster homes for the first 3 years of life

Time span: 3 years

Meth. of obs. and test.: 2 check lists of problem behavior

Findings

Children who spent the first 3 years of life in a home tended to have fewer neurotic symptoms; 13 of those from foster homes had no symptoms whereas no child from an institution was free of symptoms. More home-raised children suffered from "passive anxiety" and "intra-family tension," but more institutionalized children were involved in the following areas: frequency of overtly anxious and aggressive behavior, destructiveness, cruelty to children, failure regarding privacy rights, need for adult attention, fear of meeting new people, social misconduct, inability to get along in groups, hyperactivity, distractibility, and limited capacity for affective relations.

117

GOLDFARB, W. Personality Trends in a Group of Enuretic Children below the Age of Ten, *Rorschach Research Exchange,* 6:28–38 (1943)

Setting: Not stated

Subjects: 8 enuretic children; age 7 years 2 months to 10 years

Time span: Variable

Meth. of obs. and test.: Rorschach Test; case histories; psychological examinations to assess mental ability

Rorschach interpretations are validated mostly by impressions of case workers and by objective observations of each child in a test situation.

Findings

Of 8 children, 6 showed evidence of strongly aggressive behavior patterns; the total for C was high, M was absent, P's were few, A% was low, vague perceptions were apparent, W–. The others showed fear and withdrawal; the author suggests that they may have suppressed their hostility.

Author's interpretations

In young children, below 10 years of age, enuresis is more likely to be a major symptom of a general lack of inhibition in affective expression. In all cases emotional immaturity whether in terms of excessive aggressiveness or excessive self uncertainty and lack of reliance is indicated.

118

GOLDFARB, W. The Effects of Early Institutional Care on Adolescent Personality (Graphic Rorschach Data), *Child Development,* 14:213–223 (1943)

Setting: Variable

Subjects: 15 children raised in institutions for the first 3 years of life; 15 children raised in foster homes for the first 3 years of life

Time span: Variable

Meth. of obs. and test.: Rorschach Test

Findings

According to graphic Rorschach data the institutional group showed greater deviation from the normal than children raised in the home. This was indicated by adherence to "concrete" attitudes and apathy toward the environment. Probably there was more unconscious motivation controlling the behavior of the institutionalized children.

119

GOLDFARB, W. Psychological Privation in Infancy and Subsequent Adjustment, *American Journal of Orthopsychiatry,* 15:247–255 (1945)

Setting: Not stated

Subjects: 15 children raised in institutions for the first 3 years of life; 15 children raised in foster homes for the first 3 years of life; average age, 12 years 2 months

Time span: Not stated

Meth. of obs. and test.: I.Q.'s; case histories

Findings

The author indicates that the institutionalized children were less well-adjusted. This was manifested by lower I.Q., lower concept formation and abstraction (especially time), absence of normal inhibitory patterns, affect hunger, emotional imperviousness with superficial relationships, absence of normal tension and anxiety reaction, and social regression. There is also a comparison of the deprived and the rejected child; the latter state is thought to be less pathological.

120

GOLDFARB, W. The Effects of Psychological Deprivation in Infancy and Subsequent Stimulation, *American Journal of Psychiatry,* 102:18–33 (1945)

Setting: Not stated

Subjects: 15 children raised in institutions for the first 3 years of life; 15 children raised in foster homes for the first 3 years of life

Time span: 9 months

Meth. of obs. and test.: Stanford-Binet Intelligence Scale; Merrill-Palmer Scale; Cattell, Williams, McFarland, and Little Achievement Scale; McCaskill-Wellman Motor Coordination Test; Vineland Social Maturity Scale; California Behavior Rating Scale; Rorschach Test

Children raised in institutions were tested before placement in homes and six months after placement.

Findings

At the original testing, those raised in institutions were inferior in I.Q., vocabulary, and language, and they were more immature according to the Rorschach Tests; there were no differences between the groups according to the Vineland scale and McCaskill-Wellman test. At the second testing, the institution-raised children showed a marked drop in social maturity, but the rest of the data remained fairly constant. The disappearance of the "removal" noted at the first testing was explained by a change in response toward the tester and the testing materials.

121

GOODENOUGH, F. The Emotional Behavior of Young Children during Mental Tests, *Journal of Juvenile Research,* 13:204–219 (1929)

Setting: Institute of Child Welfare, University of Minnesota
Subjects: 990 children; age 18 months to 6 years
Time span: 1,897 observations, or 1–5 per child; 72 children tested 3 or more times over an interval of about 1 year
Meth. of obs. and test.: Children classified by age, sex, paternal occupation, and position in family; Kuhlmann-Binet Test; Minnesota Pre-School Scale

At the close of the test, ratings were immediately assigned to each child on the three traits of shyness, negativism, and distractibility. Each trait was rated on a five-category continuum.

Findings

Shyness — There was no sex difference regarding shyness in upper social class children. Boys of lower social class were less shy than girls of the same status and than either boys or girls of upper social class. This was thought to reflect the greater degree of freedom given these boys. Only children in general were less shy than children from families of more than one. The differences were most marked between only children and oldest children.

Negativism — On the whole boys were more negativistic than girls. Upper social class boys were much more negativistic than lower class boys and upper class girls, whereas in the lower social class, girls were slightly more negativistic than boys.

Distractibility — Only children were very much more distractible than children from families having more than one child, the difference being most marked between only and oldest children. A greater tendency to distraction was noted in boys and in the upper social class.

Results of testing — For each of the three traits, a decided improvement was seen with increasing age, starting at 18 months for girls and 30 months for boys. The sex differences were in accordance with previous findings by other investigators. On the average the 72 children tested at 9 and 43 weeks showed a high correlation between the first and second ratings, much less between the second and third, and insignificant correlation between the first and third.

Author's interpretations

"While the ratings serve as useful indications of behavior

characteristic of the individual at a given level of development, these characteristics may become greatly modified through subsequent training and experience."

122

GOODENOUGH, F., and LEAHY, A. M. The Effect of Certain Family Relationships upon the Development of Personality, *Pedagogical Seminary and Journal of Genetic Psychology,* 34:45–71 (1927)

Setting: Guidance clinic

Subjects: 322 clinic children, age 2 to 19 years; 293 kindergarten children, age 5½ to 6 years

Time span: Not stated, but at least 5 months

Meth. of obs. and test.: Records of family position; clinic case histories; teachers' ratings; intelligence tests; general observations

CLINIC CHILDREN

Classification of clinic cases showed that 30% of the cases were oldest children, 32% middle, 25% youngest, and 13% only. Descriptive reports were difficult to analyze because of the extreme variability of the subjects.

Findings

Oldest children were relatively high in most forms of misconduct. Middle children tended to show more sex misconduct, stealing, and direct social offenses. This group was lowest in negativistic attitudes, cruelty, refusal to submit to authority, nervousness, fear, and worry. Youngest children had no outstanding characteristics. Only children were high in negativism, disobedience, tantrums, nervousness, fears, feeding problems, and enuresis.

KINDERGARTEN CHILDREN

Of the 293 children, 22% were the oldest in their families, 20% were middle children, 42% were the youngest, and 16% were only children.

Findings

The oldest children showed a lack of aggressiveness and leadership abilities and were suggestible and gullible. They were apt to be seclusive and introverted and were low in self-confidence. Middle children were suggestible and gregarious, craved physical affection, and had flighty attention. The youngest children had no outstanding characteristics. Only children were more aggressive and self-confident; were highly gregarious, moody, excitable, and distractible; and craved physical affection.

Frequency of behavior extremes — The oldest children indicated 22.5% of the extreme ratings; the middle, 20.6%; the youngest, 16.5%; and only children, 19.7%. The difference between the oldest and youngest was four times the standard deviation of the differences and therefore was certainly significant.

Authors' interpretations

One may justifiably conclude that the disproportionate number of oldest children found among delinquents is not wholly an artifact and is related to position in the family.

123

GREEN, E. H. Group Play and Quarreling among Preschool Children, *Child Development,* 4:302–307 (1933)

Setting: Nursery school

Subjects: 40 children; age 2.1 to 5 years

Time span: Not stated

Meth. of obs. and test.: Time-sampling observations

Findings

Friendship indices increased regularly with age. From 2 to 3 years of age this was due to an increase in the number of friends; from 3 to 5, it was due to an increase in the depth of friendship. Friendship and quarrelsomeness indices correlated to .30. The ratio of quarreling to friendships decreased regularly with age. Sex differences were small; girls had slightly more friends than boys, but boys formed deeper friendships. Unisexual friendships predominated. Boys were more quarrelsome than girls. Quarrelsomeness depended on the extent to which boys made up the group; boy-boy groups were most quarrelsome, girl-boy groups less so, and girl-girl groups least.

Author's interpretations

Mutual friends were more quarrelsome and mutual quarrelers more friendly than the average. Therefore, quarreling is a part of friendly social intercourse at this age.

124

Hagman, E. P. *The Companionships of Preschool Children,* (Studies in Child Welfare, vol. 7, no. 4). University of Iowa, Iowa City, 1933

Setting: Preschool nursery

Subjects: 15 children, age 2; 24 children, age 4

Time span: 3 months

Meth. of obs. and test.: Time-sampling observations during free play; intelligence tests

Findings

Correlations and factors in companionships — Statistically little or no correlation was found with the frequency of companionship between two children and similarity in CA, MA, I.Q., height, weight, extraversion, social stimulus, or social reaction indices. The previous associations of children outside school were more closely related to companionship in school than the number of days spent together during the year or school associations in previous years. Individual children varied greatly; some chose companions similar to themselves with respect to a particular characteristic, while others chose very dissimilar companions. Some did not react to another child more than 4–5% of their total companionship opportunities; others reacted to one individual in over 40% of the occasions presenting such an opportunity. Results from asking children their preferences showed some relation between the choices expressed and actual companions.

Comparison of 2- and 4-year-olds — The 2-year-olds were significantly more similar to their most frequent (rather than their least frequent) companions in extraversion, social stimulus, and social reaction indices. The 4-year-olds seemed more similar to their frequent companions in the amount of social reaction.

While 4-year-olds definitely preferred their own sex, 2-year-olds showed no preferences. The 4-year-olds spent 58% of their time in companionship reactions; the 2-year-olds spent 50%.

125

HATTWICK, B. W. The Influence of Nursery School Attendance upon the Behavior and Personality of the Preschool Child, *Journal of Experimental Education,* 5:180–190 (1936)
Setting: Nursery school
Subjects: 2 groups of 106 children each
Time span: 2 years
Meth. of obs. and test.: Rating scales of 60 behavior items

Findings

Social adjustment — Children were more sociable after longer nursery attendance — i.e., avoidance of other children, fear of strangers, clinging to adults, etc. diminished. Social behavior seemed to be influenced over and above the natural improvement with age. Social interests increased equally for 3- and 4-year-olds, but certain techniques improved more at the 4-year-old level — i.e., there was less attacking and grabbing, and fewer refusals to share. The 3-year-olds who had attended nursery school showed more dependence on adults than those who were new at the school; this group seemed to adjust by putting the teacher in the role of mother. By the time the child was 4 years old his position in the group was better established.

Expressive behavior — Expressive behavior increased with nursery attendance. Fears, avoidance of risks, daydreaming, and nervous tendencies decreased, especially hair twisting, tension at rest, wiggling, and play with fingers.

Routine behavior — Eating problems, enuresis, and unfinished tasks decreased with nursery school attendance. The 4-year-olds probably dawdled less because they were maturing physically. The 3-year-olds were more influenced in the areas of refusal of food and resistance to rest. Sleeping habits seemed least affected.

Factors not affected by nursery school attendance — No relation was found between nursery school attendance and crying, jealousy, thumb-sucking, awaking crying, sulking, twitching, tension, temper outbursts, misrepresentation of facts, desires to be fed, taking property secretly, etc.

Author's interpretations

It is felt that in general nursery school attendance is more effective in establishing routine habits for 3-year-olds than 4-year-olds except where physical maturity is a factor.

126

HAVIGHURST, R. J., and TABA, H. *Adolescent Character and Personality.* Wiley, New York, 1949

Setting: "Prairie City," a midwestern community of 10,000 inhabitants

Subjects: 144 children; age 16 years

Time span: Variable

Meth. of obs. and test.: Character reputation — teachers' and employers' ratings; character sketch; "Guess Who" test. Social context — social status classification; socio-economic index; family relations questionnaire; interviews with teachers, families, and neighbors; records of membership in organizations; adult "Guess Who" test; Moreno-type sociometric test. Individual characteristics — essays by subject on life ideals; attitude questionnaires; Life Situation Problem Test; Moral Ideology Questionnaire; emotional response questionnaires; Mooney Problem Check List; California Personality Inventory; interviews with subjects and their families; I.Q. tests; school grades; Rorschach Test; Thematic Apperception Test; interest inventory

The authors present a study of character development and the factors which account for the differences and similarities in character. Character is evaluated by the criterion of reputation and the reliability of the latter is checked and felt to be useful. Character is defined here in the moral sense as that part of personality most subject to social approval; it is a product of environment and personal make-up. Statistical and case-study data are presented. They also demonstrated the influence of culture and social environment upon the development of personality. The personal characteristics which favor the development of good moral character are intellectual understanding of moral principles and a conviction that these principles are worth sacrifice. Moral character, personality, and social environment are systematically related. It is indicated that moral character cannot be usefully studied apart from the total personality. The authors emphasize that good social adjustment contributes to good character. Any young person who experiences success and security in home and school is likely to abide by the accepted code of morality.

127

HEATHERS, G. Emotional Dependence and Independence in a Physical Threat Situation, *Child Development*, 24:169–179 (1953)

Setting: Laboratory

Subjects: 56 children

Time span: A cross-sectional study done in the setting of a longitudinal one

Meth. of obs. and test.: Ratings on dependence and independence in a physical-threat situation; Fels Parent Rating Scale

Findings

Girls and boys had similar scores. The amount of dependence or independence children showed during this test was related to certain aspects of their mothers' behavior toward them. That is, child-centeredness (household revolves around child) was related to dependency; high acceleration (child is deliberately trained to develop skills) was related to independence.

128

HENKE, M. W., and KUHLEN, R. G. Changes in Social Adjustment in a Summer Camp; a Preliminary Report, *Journal of Psychology*, 15:223–231 (1943)

Setting: Summer camp
Subjects: 163 boys; age 8 to 18 years
Time span: 1 camp season
Meth. of obs. and test.: Washburne Social-Adjustment Inventory; Rogers Test of Personality Adjustment

Findings

The 70 older boys (mean age, 14.3) gained 5.14 points (CR 2.1) on the total Washburne inventory and showed gains in "happiness" and "impulse judgement." "Alienation" scores (group membership feelings) and other subtest scores did not change materially. The 93 younger boys (mean age, 10.4) showed no reliable changes on the total Rogers test or on subtest scores. Of 18 underprivileged boys, 15 showed reliable losses in social adjustment as indicated by the Rogers test.

129

HERTZ, M. R., and BAKER, E. Personality Changes in Adolescence, *Rorschach Research Exchange,* 5:30 (1941)
Setting: Not stated
Subjects: 76 adolescents
Time span: 3 years
Meth. of obs. and test.: Rorschach Test

Findings

Occurrence of personality traits — Traits such as inner living, responsiveness to the environment, and total amount of psychic energy available showed a moderate degree of stability. However, significant personality changes occurred between 12 and 15 years and characteristic patterns were identified for the groups at these ages.

Age 12 years — The 12-year-old tended to engage less in inner living and fantasy life, and responded more readily to the environment. He was apt to be livelier, more labile than stable, more excitable than controlled, and more "outward" in emotional behavior. Extratensive elements dominated the personality.

Age 15 years — The 15-year-old tended to withdraw within

himself; general responsiveness to the environment decreased. He was more creative. Fewer infantile patterns appeared with less egocentricity and impulsiveness and fewer outbursts. There was a definite increase in emotional adaptability and stability.

Authors' interpretations

The authors conclude that there are profound personality changes from age 12 to 15, from extratensiveness to introversiveness, with an attendant increase in adaptability and stability.

130

HILDRETH, G. Three Gifted Children; a Developmental Study, *Journal of Genetic Psychology,* 85:239–262 (1954)

Setting: School

Subjects: 4 boys; 3 gifted, 1 average

Time span: 7 years

Meth. of obs. and test.: Measurements of height and weight; Stanford-Binet Intelligence Scale; Goodenough "Draw-a-Man" test; Vineland Social Maturity Scale; manual dexterity test; Hildreth Personality and Interest Inventory; science achievement tests; character sketches

The gifted children were of the same size as average children of similar social class, but had slightly poorer health. The gifted children's relative I.Q.'s were constant but showed more of an increase with age than those of normal children. Bright children were superior on the Vineland Social Maturity Scale and manual dexterity tests, had wider interests, and read more books. The higher the I.Q.'s, the more precarious the personalities. However, some of the differences in personality might have been explained on the basis of differences in the homes of the children.

131

HIRSCH, N. D. An Experimental Study upon Three Hundred School Children over a Six-Year Period, *Genetic Psychology Monographs,* 7:487–549 (1930)

Setting: Schools

Subjects: 300 school children; age 6 years 10 months to 8 years at first testing
Time span: 5 years
Meth. of obs. and test.: Otis Primary Test; Otis Advanced Test; Woodworth-Mathews test; data on economic, anthropological, and sociological levels; anthropometric measurements

Not all of the data were obtained for all subjects.

Findings

Changes and deviations — The average in I.Q. points for all subjects on all tests was 5.26. The average deviation in mental testing was 5.3 for 173 boys and 5.1 for 120 girls; this is not a significant deviation. There were greater fluctuations among children with superior I.Q.'s. These fluctuations were not accounted for by emotional instability and were probably due to a greater variability of interest and effort and perhaps greater irregularities in the general I.Q. growth.

Correlations — There was no correlation between the degree of constancy of the I.Q. and the interval of time between tests, provided the interval was at least a few months and not more than 3 years. Comparing Otis (I.Q.) and Woodworth-Mathews (emotional instability) scores, a +.25 correlation was found between intellectual and emotional stability for boys and a +.14 correlation for girls — i.e., the greater the child's intelligence, the greater his emotional stability. Other studies confirmed these findings. There was a .5 correlation between parental economic status and the child's I.Q. Comparisons between I.Q. and cephalic index revealed a small positive correlation. In general, the larger the family (up to 6 children), the lower the I.Q.'s.

132

HOEFER, C., and HARDY, M. C. Later Development of Breast Fed and Artificially Fed Infants, *Journal of the American Medical Association,* 92:615–619 (1929)
Setting: Schools
Subjects: 383 children; age 7 to 13 years
Time span: 5 years

Meth. of obs. and test.: "Home blank" filled out by parents; medical histories and examinations; developmental records; interviews with mothers in the home; anthropometric data; educational and psychological examinations

Findings

Influence of type of feeding — On the whole children who were artificially fed were inferior physically and mentally to those who were breast fed. Except for height, the former were ranked lowest in all the physical traits measured. For example, they were the poorest nourished group, were more susceptible to childhood diseases on the average, were slowest in talking and walking, and had the smallest percentage of I.Q.'s of 120 or more. No artificially fed child was classified as exceptionally bright (I.Q. of 130).

Influence of length of breast feeding — Children who were breast fed from 4 to 9 months were definitely superior physically and mentally to all other groups. Those fed exclusively by breast for over 9 months had the lowest I.Q.'s of all groups although they apparently developed physically at a fairly normal rate.

133

HOLWAY, A. R. Early Self-Regulation of Infants and Later Behavior in Play Interviews, *American Journal of Orthopsychiatry,* 19:612–623 (1949)

Setting: Nursery school

Subjects: 17 children

Time span: Experimental behavior correlated with earlier longitudinal study

Meth. of obs. and test.: Controlled doll play; interviews; observations through one-way mirror; detailed records from longitudinal data of development of feeding and toilet procedures; rating scales

The child's behavior was scored by two psychologists as realistic, fantastic, hostile, or tangential and compared with data on feeding and toilet training in an attempt to determine the relationship between play behavior and early self-regulation in toilet training and feeding.

Findings

Self-regulated children showed more reality play, whereas strictly trained children showed more fantasy play. Pooled ratings of these children by the nursery staff corroborated the clinical judgment that realistic play in the doll situation indicates a good adjustment and social spontaneity.

Author's interpretations

The author believes that the satisfaction of individual needs is more important than the time or type of training procedure. She also believes that the best indication of a lack of frustration in the child's early life is his present ability to face his home situation realistically.

134

Hopkins, C. D., and Haines, A. R. A Study of One Hundred Problem Children for Whom Foster Care was Advised, *American Journal of Orthopsychiatry,* 1:107–128 (1931)

Setting: Institute for Juvenile Research in Chicago, Ill.

Subjects: 100 foster children with behavior problems; age 1 to 17 years; mean age, 10 years

Time span: Not stated

Meth. of obs. and test.: Social work case histories

Findings

The children in this study came from "impossible homes." It was not feasible to divide them into problem and nonproblem groups, but it was obvious that in nearly every case the home environment had produced pathological effects. Of those placed in foster homes as recommended, 72% showed successful adjustment; this corresponds to the usual rate (65–75%) of successful adjustment in foster homes, regardless of the number of individual problems. Of the children who were not placed in foster homes only 9% made successful adjustment.

135

Hulson, E. L. An Analysis of the Free Play of Ten Four-Year-

Old Children through Consecutive Observations, *Journal of Juvenile Research,* 14:188–208 (1930)

Setting: Nursery school

Subjects: 10 children; age 4 years

Time span: 4½ to 9 months

Meth. of obs. and test.: Detailed, consecutive, 5-minute observations during free play period; data gathered on choice of material, length of interest, persistence of choice, order of choice, and child's reactions

Findings

Choice of materials — Blocks ranked first in the categories of the number of times chosen, number of minutes used, persistence, and social value. Sand, watching, house corner, Kiddy-kar, and see-saw ranked high, while blackboard, animals, and dolls ranked consistently low. House corner, sand, blocks, see-saw, dishes, and Kiddy-kar were the materials most often involved in activities. The material of greatest appeal tended to be chosen first. If a material was chosen often, then the time spent and day-to-day persistence in its use were also relatively high. Wide divergencies were found in the amount of time spent on a choice. Interest span decreased from choice to choice. The mean length of time spent on first choice varied from 74 to 268 minutes.

Watching — Watching was part of the activity of 9 of the 10 children. There were four types found — watching the general situation before making a choice (44% of time); watching one material and choosing another, sometimes because of an inadequate supply (29%); watching a material and then choosing it (21%); and watching aimlessly and dreamily (5%).

136

HURLOCK, E. B., and SENDER, S. The "Negative Phase" in Relation to the Behavior of Pubescent Girls, *Child Development,* 1:325–340 (1930)

Setting: Schools and court

Subjects: 142 girls referred to children's court and compared with questionnaire data
Time span: Variable
Meth. of obs. and test.: Questionnaires; family, developmental, and social histories; physical and psychological examinations

Findings

Negative traits — The most frequent negative traits in pubescent girls were restlessness, instability, loss of school interest and friends, desire to be alone, neglect of creative activity, withdrawal from companions, tendency to be destructive, loss of interest in work, withdrawal from leadership, and hostility toward society.

Environmental factors — There was a wide disparity with respect to negative traits between girls from poor homes and those from better homes. Girls from poor homes accounted for 80% of the total hostility to society and 71% of the loss of interest in school and friends. They were also preponderant in their desire to be alone, their withdrawal from friends, their loss of interest in work and leadership, and their instability. Restlessness was evenly distributed between the two groups. In general the girls showing "negative phase" traits belonged to inferior socio-economic levels, had poor home conditions, were mentally retarded, and lived mainly in foreign and congested areas in New York.

Court cases — Most of these cases came from poor congested areas; 65% had foreign parents; 58% either had no father or mother, or unmarried parents. Delinquents were usually of borderline intelligence. There was no preponderance of sexual delinquency in this group in the thirteenth year (as had been found by C. Bühler) but relatively more at age 13–14 than at age 15.

Authors' interpretations

Hetzer's conclusion that "negative phase" traits are concomitant with puberty is unwarranted. The occurrence of sexual delinquency also depends on many factors; the drive is reinforced rather than introduced at puberty. Mental retardation seems to prevent sublimation of sexual desire. Restlessness and instability may be concomitants of female puberty, but the degree depends on many factors, especially environment and home training. But neither environment nor mental caliber are as significant in behavior as are the forces of individual personality.

137

Jackson, E. B., and Klatskin, E. H. Rooming-in Research Project; Development of Methodology of Parent-Child Relationship Study in a Clinical Setting, in: *Psychoanalytic Study of the Child* (vol. V). International Universities Press, New York, 1950

Setting: Rooming-in service, home and clinic follow-ups, nursery school

Subjects: Not stated

Time span: 3-year survey and continued observation of some of the subjects at the Child Study Center Nursery School

Meth. of obs. and test.: Prenatal interviews by pediatrician; subjective impressions of labor; notes by doctors and nurses; home visits; pediatric examinations; Cattell Infant Intelligence Scale; Rorschach Test; observations of doll play

This report attempts to describe the rooming-in service and evaluate its effect. A case study is included.

138

Jackson, E. B., Klatskin, E. H., and Wilkin, L. C. Early Child Development in Relation to Degree of Flexibility of Maternal Attitude, in: *Psychoanalytic Study of the Child* (vol. VII). International Universities Press, New York, 1952

Setting: Rooming-in service, home, and clinic

Subjects: 3 primiparous mothers and their infant daughters

Time span: Not specifically stated, but at least through the first year of life

Meth. of obs. and test.: Prenatal interviews by pediatrician; subjective impressions of labor; notes by doctors and nurses; home visits; pediatric examinations; Cattell Infant Intelligence Scale; Rorschach Test; observations of doll play

These three case histories contrasting attitudes of overpermissiveness, moderation, and rigidity indicated that the basis for the divergence in the attitude of the mothers studied (and by inference other mothers) is to be found in their own experience and interpersonal relationships. Although the three mothers had been exposed to the same philosophy of child care, the records show that each applied it differently in practice. The histories indicated that either insistent permissiveness or extreme rigidity exerts an adverse effect upon infants and young children.

Authors' interpretations

The authors are inclined to believe that the ultrapermissive mother may be more readily susceptible to change through therapeutic influence and cultural pressures than the ultrarigid mother. They point out that both extremes are rooted in ambivalence.

139

JACOBSON, W. A Study of Personality Development in a High School Girl, *Rorschach Research Exchange*, 2:23–35 (1937)

Setting: Not stated

Subject: 1 girl; age 17 years

Time span: Several months

Meth. of obs. and test.: Case history; Rorschach Test; Bernreuter Personality Inventory; Strong Vocational Interest Blank for Women; Terman Group Test of Mental Ability; classmates' personality evaluations; observations by teachers, principal, and employer

This article shows how the Rorschach Test can serve as an instrument for measuring personality development in high school girls.

Findings

Rose, the subject, was a girl who showed a lack of maturity in her emotional life. During the short period of time she was studied, improvement could be noted in that she overcame to a marked degree the constrictive attitude shown in her first Rorschach performance (M). General growth was indicated by the increases in all personality factors, with greater freedom (FM), greater awareness of inner troubles (FK) and of the external world (c), and greater readiness to make adjustments to these external forces (FC). The Rorschach data substantiated and corroborated findings from other sources, i.e., personality test and other people's evaluations.

140

JERSILD, A. T. The Constancy of Certain Behavior Patterns in Young Children, *American Journal of Psychology,* 45:125–129 (1933)

Setting: Nursery school

Subjects: 11 children

Time span: 1 year

Meth. of obs. and test.: Timed records of overt activity; use of objects; contacts with other children; laughing and talking

Findings

The children were tested for several weeks on two occasions, one year apart. The quantitative records showed a rather low correlation between the two tests for individual children. Several children retained some of their nursery school characteristics; one child's conduct changed markedly.

Author's interpretations

The study apparently gave inadequate indications of basic behavior tendencies. The author feels that records of the frequency and duration of a child's activities are not sufficient and suggests that additional measures of content and scope are necessary.

141

JERSILD, A. T., and FITE, M. D. *The Influence of Nursery School Experience on Children's Social Adjustments* (Child Development Monographs, no. 25). Teachers College, Columbia University, New York, 1939

Setting: Nursery school

Subjects: 18 children; age 3 years

Time span: School year plus previously obtained data

Meth. of obs. and test.: Modified diary of school behavior; 10 15-minute observations and 8 5-minute observations of each child; case records; school reports; scores of social contacts

Findings

General trends of social behavior — The structure of the group was the most important factor in determining social behavior. The general trend was toward a sharply increased amount of fraternizing though not all followed this trend. The year's experience probably reduced by little the individual differences in social contacts, save those associated with previous nursery schooling. Children's adjustments and the effect of nursery school environment were highly individual matters. The trend of group results did not do justice to the individual cases.

New relationships — Three-year-olds with previous nursery school experience were initially more receptive to social contacts than new children. However, this sociability derived from nursery attendance was not so much increased ability and inclination as a carry-over of special companionships. The new children made large initial gains and by spring they had, as a group, overcome their initial disadvantage.

Authors' interpretations

It is necessary to interpret the child's various contacts in relation to the record of his behavior as a whole. It is the writers' opinion that opportunities to encourage skills and aptitudes are too frequently neglected in preschool education.

142

Jersild, A. T., and Markey, F. V. *Conflicts between Preschool Children* (Child Development Monographs, no. 21). Teachers College, Columbia University, New York, 1935

Setting: Nursery school

Subjects: 54 children, age 22 to 50 months; 36 studied longitudinally

Time span: 1 year

Meth. of obs. and test.: Records of conflicts in free play, using 15-minute observations; teachers' ratings

Conflict was defined as any instance of, or attempt to, attack (by word, deed, or gesture) or interfere with another person, his activities, or possessions.

Findings

Characteristics of conflicts — On the basis of timed observations, there was a total of 1,577 conflicts averaging 30.9 (SD 18.7) per child during 150 minutes, or 1 every 5 minutes. On the basis of additional indirect observations, the average was 58 (SD 29.1). The median duration was between 20 and 30 seconds. Overt or verbal aggression against another's material possessions accounted for two-thirds of the conflicts; 23 instances were classed as sympathetic, or 3 times to each 200 conflicts. Preschool conflicts were usually transitory and rarely were they due to revenge, alliance, chronic feelings, etc. There was a wide range of individual behavior; for example, one child had 141 conflicts while another had 17. There was a high degree of constancy in the individual child's behavior with an overall correlation of .79. In terms of absolute frequencies there were positive correlations among practically all aspects of combative behavior, and between the frequency of aggression and the frequency of being a victim of aggression.

Factors in conflicts — There was an irregular and inconsistent decline in the frequency of conflict with age. They were more frequent between two older or two younger children than between an older and younger child. Without regard to age, boys had more conflicts than girls; considering age, results showed more conflicts for the sex in the majority. All sex differences were small and unreliable, though they were generally more apparent with an increase in age. Same-sex conflicts were more frequent than opposite-sex conflicts. There were low correlations between conflicts and height, weight, and their ratios. Unreliable differences were found in the influence of nationality upon conflicts; Southern Europeans had more conflicts than Northern Europeans who in turn had more than Jews. There was a slight negative correlation between the frequency of conflicts and I.Q., and a consistent positive correlation between I.Q. and the frequency of verbal conflicts.

Increase in conflicts — In the second year of nursery school, there were more conflicts per child, more aggression, more verbal attacks, more personal attacks, and fewer victims. These increases were ascribed to nursery attendance, small space, and the learning of aggressive acts. There was a marked increase in conflicts accompanied by an increase in language skills and decrease in crying. The former day nursery members showed a decline in the frequency of conflicts.

Teacher as factor in conflicts — Teachers interfered in 32% of the total number of occurrences. In 72% of the interferences the teacher decided the issue in favor of one or the other contestant, usually the victim. The teacher tended to side with a child who lost many fights when left alone and against a child who usually won alone. Observers recorded 130 conflicts directed at teachers. There was a positive correlation between the frequency of conflicts with other children and with teachers. Teacher ratings (composite) correlated .53 with the children's actual scores.

143

Jones, H. E. Physical Ability as a Factor in Social Adjustment in Adolescence, *Journal of Educational Research,* 40:287–301 (1946)

Setting: Research institute

Subjects: 78 boys; 10 subjects were selected from each end of the distribution scale of total strength scores and then compared

Time span: 6 to 8 years

Meth. of obs. and test.: Anthropometric measurements; x-rays; medical records; athletic performance; reputation test; rating and observations from longitudinal data; personal-social inventory (based on Rogers Test of Personality Adjustment)

Findings

Strong boys — Of the 10 strong boys (as determined by composite scores of measures of grip, pulling strength, and thrusting strength), 9 were above average in height and weight. The average body build based on Sheldon Somatotyping was 2½–5–

3½. The group was above average in skeletal maturation, health records, and athletic abilities. The members rated high in popularity and social prestige and were well adjusted.

Weak boys — These 10 boys showed pronounced tendencies toward asthenic physique, later maturation, poor health, social difficulties and lack of status, inferiority feelings, and personal maladjustment in other areas. The average body build was 2½–3½–4½.

Social ratings — The reciprocal action of these factors as well as hereditary and environmental agencies was considered. The social ratings for the strong group were consistently favorable and showed an apparent upward trend in popularity. The weak boys showed somewhat marked changes in an unfavorable direction over a 5-year period. These observational findings were paralleled for the same cases and for the same intervals by adjustment scores. In the early record the strong group had scores of 55 while the weak group had 48. But 5 years later the differences were more marked and it was thought that they would have been still greater were it not for the weak boys' use of denial.

144

JONES, H. E., *et al.* *Development in Adolescence; Approaches to the Study of the Individual.* Appleton-Century, New York, 1943

Setting: Research institute

Subject: 1 boy used as an example of the longitudinal data

Time span: 7 years

Meth. of obs. and test.: Anthropometric measurements; x-rays; medical records; athletic performance; reputation tests; rating and observations from longitudinal data; personal-social inventory (based on Rogers Test of Personality Adjustment)

The subject, a young boy, is presented in many aspects. His physical and social environment are depicted and his adolescent character described.

Findings

The boy had a stressful adolescence with a low point at age

15 when his reputation among classmates was poor. Adults rated him as anxious, and his drive ratings indicated he was in marked conflict. These psychological factors were correlated with a delay in the adolescent physical growth spurt and family interaction with an overprotective mother. He is shown overcoming his conflicts and developing a sense of humor and a social role as he matured.

145

JONES, M. C. Guiding the Adolescent, *Progressive Education,* 15:605–609 (1938)

Setting: Research institute

Subjects: Not stated

Time span: 7 years

Meth. of obs. and test.: Anthropometric measurements of height, weight, etc.; physiological measurements; motor tests. Intelligence and achievement tests — Terman Group Test of Mental Ability; Kuhlmann-Anderson Intelligence Test; Thorndike CAVD; Stanford-Binet Intelligence Scale; Stanford Achievement Test; intellectual reports. Emotional characteristics and adjustment — autonomic system; self-reports; interviews; Thematic Apperception Test; Rorschach Test; self-portraits; voice records, etc. Social behavior — observations by staff; reputation test; interviews with parents. Description of methods for observation of social behavior — (1) Clubhouse provided a chance to see much social interchange. Records were made after the time of observation and when possible two observers made records at the same time. Records were verbatim accounts, narrative records, interpretative comments and specific ratings. Ratings were of expressive characteristics, social orientation, and social status. (2) Dances, week-end excursions, etc. (3) Observations during the children's day at the clinic

Findings

The principal conflicts of adolescents involve adjustment to the opposite sex and disputes with parents over grades, money, and group activities. Girls usually mature 2 years earlier than boys, though either may lag 4 to 5 years behind the group in the development of appropriate heterosexual patterns of behavior. A third area of conflict arises around the attempt at development of individual integrity and independence.

Author's interpretations

The author comments that adolescents are slavishly submissive to authority rather than rebellious against it. However, they have substituted the authority of their peers for the authority of adults. Dependence on the group provides a transition between dependence on parents and personal maturity and thus represents a phase of growth. At first adolescents tend to sample various groups and to experiment with friendships; later they evolve tight barriers around small groups to provide security and assured acceptance. Society and parents should encourage self-confidence through creativity and the realization of abilities. Guidance should furnish the adolescent with opportunities for making decisions, achieving status among his associates, and building individual patterns of behavior which will lead toward a mature personality.

146

JONES, M. C., and BAYLEY, N. Physical Maturing among Boys as Related to Behavior, *Journal of Educational Psychology*, 41:129–148 (1950)

Setting: Research institute

Subjects: 2 groups of 16 boys who fall at opposite ends of a normal distribution curve on assessments of skeletal age

Time span: Not stated

Meth. of obs. and test.: Anthropometric measurements of height, weight, etc.; physiological measurements; motor tests. Intelligence and achievement tests — Terman Group Test of Mental Ability; Kuhlmann-Anderson Intelligence Test; Thorndike CAVD; Stanford-Binet Intelligence Scale; Stanford Achievement Test; intellectual reports. Emotional characteristics and adjustment — autonomic system; self-reports; interviews;

Thematic Apperception Test; Rorschach Test; self-portraits; voice records, etc. Social behavior — observations by staff; reputation test; interviews with parents. Description of methods for observation of social behavior — (1) Clubhouse provided a chance to see much social interchange. Records were made after the time of observation and when possible two observers made records at the same time. Records were verbatim accounts, narrative records, interpretative comments and specific ratings. Ratings were of expressive characteristics, social orientation, and social status. (2) Dances, week-end excursions, etc. (3) Observations during the children's day at the clinic

Findings

Comparison of early- and late-maturing boys — Boys maturing early were consistently rated superior in physical attractiveness, had better personal habits, and were more relaxed. Boys maturing late were rated lower in attractiveness, were above average in expressiveness and attention-seeking, and were uninhibited but tense. There was no marked difference in popularity, leadership, prestige, poise, assurance, cheerfulness, or social effect on the group.

Comparison of early- and late-maturing girls — There were great differences in behavior in girls. Those maturing early were better in social interchange and seemed to have somewhat more social status and less need to strive for it. The girls maturing late were more intent upon getting attention, more restless, more assured in class, more talkative and bossy, less grown-up, less good-looking, and less apt to have older friends.

147

JONES, M. C., BAYLEY, N., and JONES, H. E. Physical Maturing among Boys as Related to Behavior, *American Psychologist,* 3:264 (1948)

Setting: Research institute

Subjects: 2 groups of 16 boys who fall at opposite ends of a normal distribution curve on assessments of skeletal age

Time span: 8 years

Meth. of obs. and test.: Anthropometric measurements of height, weight, etc.; physiological measurements; motor tests. Intelligence and achievement tests — Terman Group Test of Mental Ability; Kuhlmann-Anderson Intelligence Test; Thorndike CAVD; Stanford-Binet Intelligence Scale; Stanford Achievement Test; intellectual reports. Emotional characteristics and adjustment — autonomic system; self-reports; interviews; Thematic Apperception Test; Rorschach Test; self-portraits; voice records, etc. Social behavior — observations by staff; reputation test; interviews with parents. Description of methods for observation of social behavior — (1) Clubhouse provided a chance to see much social interchange. Records were made after the time of observation and when possible two observers made records at the same time. Records were verbatim accounts, narrative records, interpretative comments and specific ratings. Ratings were of expressive characteristics, social orientation, and social status. (2) Dances, week-end excursions, etc. (3) Observations during the children's day at the clinic

Case material is presented.

Findings

The greatest contrasts in physical characteristics occurred between the ages of 13 and 15, whereas the greatest psychological differences took place between the ages of 15 and 16. During the former period those with early osseous maturation were masculine in build while those maturing late were childish in build. Boys with accelerated skeletal development were considered more mature and attractive by adults and classmates and were accorded more status which they did not strive for particularly. From their ranks came outstanding student leaders. The boys with retarded skeletal development exhibited relatively immature behavior and reacted to their temporary physical disadvantage by greater striving for attention or by withdrawal.

148

Kanner, L. Problems of Nosology and Psychodynamics of Early Infantile Autism, *American Journal of Orthopsychiatry*, 19:416–426 (1949)

Setting: Hospital

Subjects: 55 children under 2 years of age who show features characteristic of the condition the author refers to as "early infantile autism"

Time span: 6 years

Meth. of obs. and test.: Psychiatric and medical examinations; developmental data

Findings

Characteristics of early infantile autism — The children in this group experienced a profound withdrawal from contact with people; an obsessive desire for the preservation of sameness; a skillful, even affectionate relationship to objects; a retention of pensive physiognomy; and mutism or language having no value for interpersonal communication. None of the subjects had any trauma or illness to which the symptoms could be ascribed, and only one child had a family history of serious mental disease.

Characteristics of parents — Almost all adult relatives had been successful in their careers. These children's parents had the following characteristics in common: high intelligence, sophistication, "mechanization of human relationships," i.e., they were serious, undemonstrative, uncomfortable with people, formal, polite, respectful, and loyal to the marital partner. In each case a lack of warmth and an obsessive devotion to duty were apparent in their behavior toward the child.

Author's interpretations

The author differentiates this condition from Heller's disease and aphasic disorders and correlates its features with the essential features of schizophrenia. Autism may be looked upon as the earliest possible manifestation of childhood schizophrenia.

149

Kasanin, J. The Affective Psychoses in Children, *American Journal of Psychiatry,* 10:897–926 (1931)

Setting: Hospital

Subjects: 10 children

Time span: 2 or more years

Meth. of obs. and test.: Psychiatric and medical examinations; family histories; psychotherapy; intelligence tests

Findings

Defining affective psychoses — In 2 cases kinetic features (restlessness, excitability, irritability) were prominent, plus motor disturbances in early infancy. In the other 8 cases mood disturbances (elation and depression) were essential elements of the clinical picture. Half the children showed definite physical anomalies. The occurrence of affective disorder in childhood carried with it a serious prognosis.

Factors influencing psychoses — The immediate precipitating factors in psychoses were usually quite trivial but dynamic and significant. The child's constitution seemed the most important factor in the etiology of the breakdowns. In 3 cases a conflict over sex was of direct etiological significance.

Results of study — Of the total, 2 children remained chronic invalids and used their illness to secure recognition and attention; one child definitely deteriorated intellectually. The latter's I.Q. dropped and he was sent to a school for the feeble-minded. The rest were functioning on an inferior level; some of these were still patients at various state hospitals and lacked curiosity, lost spontaneity, and had poor social adjustment.

Author's interpretations

Affective psychoses are rare in childhood (2 out of 1,900 admissions at this hospital). However, this infrequency is partially due to a lack of recognition and to the attribution of symptoms to schizophrenia or physical illness.

150

Kasanin, J., and Veo, L. The Early Recognition of Mental Dis-

eases in Children, *American Journal of Orthopsychiatry,* 1:406–429 (1931)
Setting: Boston Psychopathic Hospital
Subjects: 8 children
Time span: Variable
Meth. of obs. and test.: Psychiatric interviews; social work case histories

The eight children in this series were studied because of extremely unfavorable constitutional factors. In every case (and especially in those later diagnosed as schizophrenics) psychological tests revealed evidence of mental disorder long before clinical symptoms became apparent. The authors believe that psychological tests offer very sensitive criteria for the early evaluation of mental health.

151

KAWIN, E. *Children of Preschool Age; Studies in Socio-economic Status, Social Adjustment and Mental Ability, with Illustrative Cases.* University of Chicago, Chicago, 1934
Setting: Institute for Juvenile Research in Chicago, Ill.
Subjects: Variable; see below
Time span: Variable
Meth. of obs. and test.: Psychiatric case histories; Merrill-Palmer Scale; intelligence tests

In this book the Institute for Juvenile Research describes its set-up for study and treatment and presents three projects.

COMPARISON OF HIGH AND LOW ECONOMIC GROUPS ON MERRILL-PALMER SCALE

Findings

High economic groups scored higher on the verbal test whereas the lower economic group scored higher on motor tests.

SOCIAL ADJUSTMENT BASED ON STATISTICAL ANALYSIS OF CASE RECORDS

Adjustment was based on a child's relationship to other children, and an attempt was made to relate specific factors in his record to social adjustment. The 635 cases were divided into a problem group, a well-adjusted group, and an unselected group. The problem group was divided according to ascendant behavior, submissive behavior, and unclassifiable behavior.

Findings

Problem children were slightly older, slightly less healthy, and had lower I.Q.'s They had the same number of siblings, but included a larger proportion of oldest children and were more involved in sibling rivalry. They were more apt to have absent or dead fathers or more negative fathers, experienced more parental disagreement on child-rearing practices, and more parental separation and divorce. Many other personality problems were more frequent in the unadjusted group; negativism, enuresis, and overdependence were more frequent, but thumb-sucking, nail-biting, and masturbation were less frequent.

ANALYSIS AND COMPARISON OF TESTS FOR PRESCHOOL CHILDREN

152

KAWIN, E. Implications of Individual Differences at the First Grade Level, *American Journal of Orthopsychiatry,* 8:654–672 (1938)

Setting: Schools

Subjects: 535 children

Time span: 3 to 4 years

Meth. of obs. and test.: Metropolitan Readiness Test; group intelligence tests; standard achievement tests; family data; teachers' ratings; progressive achievement tests; Kuhlmann-Anderson Intelligence Test

Findings

A marked variability of development in individual children was found when the four factors (CA, MA, I.Q., and first grade readiness) were used. Follow-ups showed that the Metropolitan Readiness Test was the most predictive of adjustment in primary grades. There was no correlation between chronological age on entering first grade and later adjustment. According to teachers' ratings, there was very little or no relation among these four factors and subsequent emotional or social adjustment.

Author's interpretations

The Metropolitan Readiness Test was quite successful in predicting first grade success, but it could not be relied upon for judging facility of learning fundamental skills or further grade progress. The author emphasizes that individual differences in development and readiness in first grade indicate the need for individualization of the school program and careful guidance.

153

KAWIN, E., and HOEFER, C. *A Comparative Study of Nursery-School versus a Non-Nursery-School Group.* University of Chicago, Chicago, 1931

Setting: Nursery school

Subjects: 22 beginners in nursery school; 22 matched controls

Time span: 1 school year

Meth. of obs. and test.: Merrill-Palmer Scale; anthropometric data; medical examinations; data on habits

Findings

With testing equal for both groups, the results on the Merrill-Palmer Scale indicated that both groups apparently gained in mental growth between fall and spring. Physical growth data (measurement of 14 physical traits) revealed no differences in development between the two groups. Medical ratings showed that in the spring both groups had poorer ratings than in the fall with the nursery group in slightly better condition, although the difference was not significant. When all the habits were totaled and scored, a larger percentage of nursery school children showed improvement in habit status — i.e., more desirable and fewer undesirable habits (especially those indicating less dependence on and greater emancipation from adults).

154

KENDERDINE, M. Laughter in the Pre-School Child, *Child Development*, 2:228–230 (1931)

Setting: Nursery school

Subjects: 26 children; age 2 to 4 years

Time span: 5 months

Meth. of obs. and test.: Records of laughter for 1 hour 5 times a week; toy experiment

Findings

Factors in laughing — Children laughed most frequently at situations involving motion of self, then situations they realized were socially unacceptable, and finally in humorous situations. The presence of other children was essential for making children laugh and those with higher I.Q.'s laughed more frequently. The average frequency of laughs for the period studied was 14 for the 2-year-olds, 6.2 for the 3-year-olds, and 6.5 for the 4-year-olds.

Differences in laughter due to age — The 2-year-olds laughed most at motion of self, socially unacceptable situations, and pleasure in occupation and accomplishment, while the 3-year-olds laughed most at socially unacceptable situations, make-believe, situations showing appreciation of humor, and motion of self. The 4-year-olds laughed at motion of self, socially unacceptable situations, and situations showing appreciation of humor.

155

KINDER, E., and RUTHERFORD, E. J. Social Adjustment of Retarded Children; a Follow-up Study of Retarded Children Seen in Henry Phipps Psychiatric Dispensary between January and June, 1921, *Mental Hygiene*, 11:811–833 (1927)

Setting: Henry Phipps Psychiatric Dispensary

Subjects: 68 children; 35 dull (I.Q. 76–90), 25 retarded (I.Q. 50–75), and 8 defective (I.Q. below 50)

Time span: 5 years

Meth. of obs. and test.: Psychiatric studies of character traits and neurotic tendencies; school histories; Revised Stanford-Binet

Intelligence Scale; medical examinations; reports from agencies, institutions, and parents

Findings

General results — Of the entire group, 40 children were in institutions or were making poor adjustments. Only 19 of the 35 in the dull group were self-supporting or gave promise of being so, 14 had drifted into institutions, and 2 were making unsatisfactory adjustments. In the retarded group the boys were doing as well as in the dull group, but relatively more girls were in institutions. No marked correlation was found between retardation and the degree of social adaptation except among 8 of the 13 girls of the retarded group who went to custodial institutions. Two individuals adjusted better than the original test findings indicated. This discrepancy suggested that in these cases emotional or circumstantial factors at the time of the test gave an incorrect estimation of ability.

Correlation of environment and social adjustment — There was a significant correlation between environment and social adjustment. For example, all but 2 of the 14 children who made satisfactory adjustment had been placed in an improved environment; of the 36 children who required institutional care, only 2 had been offered an improved environment for placement.

Authors' interpretations

Data, such as the finding that all but one who made satisfactory or fairly satisfactory adjustment reached at least the fifth grade in school, suggest that some retarded children can benefit from training adapted to their needs.

156

KLATSKIN, E. H. Shifts in Child Care Practices in Three Social Classes under an Infant Care Program of Flexible Methodology, *American Journal of Orthopsychiatry*, 22:52–61 (1952)

Setting: Clinic and home

Subjects: 249 mothers from 3 social classes

Time span: 1 year

Meth. of obs. and test.: Minimum of 4 prenatal interviews or 2 postnatal visits at home; parental replies to questionnaires with evaluations of child behavior and child-care practices; monthly evaluations of a few subjects; social service evaluations

Findings

Results showed an increased leniency in mothers who participated in the program regardless of social class.

157

KLOPFER, B., and MARGULIES, H. Rorschach Reactions in Early Childhood, *Rorschach Research Exchange*, 5:1–23 (1941)

Setting: Not stated

Subjects: 155 children; age 2 to 7 years

Time span: 5 years (variable)

Meth. of obs. and test.: Rorschach Test

Two cases are presented — that of a slightly retarded boy from the age of 2 years 11 months to 4 years (4 records) and a well-adjusted girl from the age of 2 years to 3 years 9 months (3 records and case history).

Findings

Young children present certain basic limitations such as co-operation and language difficulties. Performance lasts about 10 minutes and requires the child's "spur of the moment" motivation. Children can generally identify their own responses during inquiry. Successive patterns of early reaction are called "magic repetition," e.g., everything in all the cards is "doggie." This response is rare after age 3. Rejection by "I don't know" or "You tell me" may be repeated or there may be partial perseveration, although both are rare after 5 years of age. The majority of the cards (7 out of 10) receive individual attention and a variety of responses. If this occurs before age 3 the child is probably very gifted and prematurely developed. If this stage has not come by age 5, the child is infantile, disturbed, or retarded.

Authors' interpretations

The authors discuss the Rorschach reactions of children in

general. They feel that the technique has "unlimited possibilities" for extensive use in longitudinal studies. Between the ages of 2 and 4 it seems worthwhile to collect a new record every 2 or 3 months and later at every half-year. "There is hardly any other means of getting a more plastic impression of mental and emotional growth in an objective form than in such longitudinal Rorschach studies."

158

KLUCKHOHN, C., and ROSENZWEIG, J. C. Two Navaho Children over a Five-Year Period, *American Journal of Orthopsychiatry,* 19:266–278 (1949)

Setting: Field trip

Subjects: 2 Navaho children studied from birth; a girl, age 6 years; a boy, age 7 years

Time span: 5 years

Meth. of obs. and test.: Rorschach Test; Thematic Apperception Test; Stewart Emotional Response Test; Bavelas Moral Ideology Test; Special Group Attitude Projective Test

By general observation and this variety of projective techniques, the authors traced the personality development of two children over a 5-year period. In general the development conformed to theoretical expectations. Changes could be traced to maturation processes and special situational determinants. Both subjects showed marked tendencies toward consistency of personality characteristics.

159

LANDER, J. Traumatic Factors in the Background of 116 Delinquent Boys, *American Journal of Orthopsychiatry*, 11:150–157 (1941)

Setting: Residential school

Subjects: 116 delinquent boys

Time span: Variable

Meth. of obs. and test.: Case histories

Findings

Traumatic factors such as parental imcompatability, emotional instability, social maladjustment, and rejection were found very frequently in the backgrounds of delinquent children. Of 116 boys in the total series, 99 had suffered from one or more of these traumatic factors; the other 17 had evidence suggestive of such trauma.

Author's interpretations

The author suggests that for prophylaxis, children from such homes must be exposed to "supplementary parents" for emotional warmth, security, and conscience building. An elaboration of kindergarten-nursery school was suggested as an attempt to meet this need.

160

LANDRETH, C. Factors Associated with Crying in Young Children in the Nursery School and Home, *Child Development,* 12:81–97 (1941)

Setting: Nursery school

Subjects: 32 children; age 2 years 8 months to 5 years 2 months

Time span: Variable

Meth. of obs. and test.: Records of crying at home and school; incident-sampling techniques

Findings

Factors related to crying — Children with high I.Q. tended to cry less especially at home. Children who slept more also tended to cry more. Boys tended to cry more than girls, especially at home. Slight colds, etc. tended to increase crying at home. The incidence of crying was related graphically to the hour of day.

Situations causing crying in nursery — An attack on a child's person accounted for 31.8% of crying for boys and 35% for girls, whereas an attack on a child's property caused crying 24.4% of the time in boys and 25% in girls. Frustration by another child caused 10.8% of the crying in boys and 14.8% in girls; frustration by an inanimate object resulted in 12.2% of boys' crying and 3.7% of girls' crying. Accidental injury caused 11.1% of the boys and 18.6% of the girls to cry. Conflict with adults caused crying in 5.4% of the boys and .8% of the girls, fear in 4.4% of the boys and 2.3% of the girls.

Analysis of causal factors — An analysis of the children's conflicts in the nursery school, which represented 75% of the causal factors of crying, revealed that boys were much more frequently associated with other children's crying than girls (ratio of 3:1). The apparent sex differences were that girls cried more often because of injury while boys cried more often because of frustration by inanimate objects and conflicts with adults. There were also indications of conflicts occurring more frequently between close friends or arising out of a marked difference in age, size, and strength of one member of the group. The type and percentage of frequency of sympathetic responses to children crying in the nursery furnished evidence of a law of diminishing returns for those who cried more frequently — the more they cried, the less sympathy they received.

Child-training as a factor — In the home, the parents' records revealed that inconsistent and poor methods of child training were responsible for the difficulties leading to crying. The individual methods of dealing with situations leading to crying varied considerably. Most frequent were ignoring, reasoning with, or spanking the child. It was usually the mother who responded to these situations.

161

LASKO, J. K. Parent Behavior toward First and Second Children, *Genetic Psychology Monographs*, 49:97–137 (1954)

Setting: Home

Subjects: 40 pairs of first and second children matched for age at the time mothers' behavior was rated

Time span: Variable

Meth. of obs. and test.: Fels Parent Rating Scale

Findings

Parental treatment — Contrary to prediction, the parents

tended to baby, protect, and be more solicitous of the second child. The second child got more warmth and less parental control. The warmth of the first child's environment was soon lost. The second child began at a less extreme level but maintained it. The oldest did get a great deal of verbal stimulation and accelerative attempts most notably at 2 and 5 years of age. The differences in treatment were most noted in the preschool level; warmth or its lack, however, was a stable characteristic of the mother in her dealings with the preschooler.

Age and sibling relationships as factors — Children displaced by siblings at 3 years of age suffered more than children displaced at age 4. The chief loss was in parent-child interaction. Children close together in chronological age benefited from more rational treatment than siblings with a wide spread between ages. After the birth of a third sibling, the mother's behavior to the second was a modified replica of her change in behavior to the first sibling.

162

LERNER, E., and MURPHY, L. B. *Methods for the Study of Personality in Young Children* (Society for Research in Child Development, Monograph, vol. 6, no. 4). National Research Council, Washington, D.C., 1941

Setting: Nursery school

Subjects: 41 children; 16 boys, 25 girls

Time span: 4 months to 3 years

Meth. of obs. and test.: Various projective techniques devised by the experimenters; Rorschach Test

Life space of the child (Kurt Lewin's concept) was used as a basis for interpretive analysis of free play with a miniature-life toy collection. The report presents no organized description of the number of times or the number of subjects exposed to this technique, but discusses sequential behavior. It describes the categories of behavior devised for classifying activity in free play situations and the techniques used for interpretations. Children who demanded to come to nursery school produced the fullest and most dynamic fantasy. To distinguish between the well-adjusted introvert and the shut-in child, a series of toys, mostly of a sensory nature, were provided. The shut-in child was identified by compulsive responses. Dough and cold cream were used to evaluate temperament and patterns of spontaneity because they were believed to be free from reality orientation. There were two main differences in the use of these substances — i.e., pleasures in handling and functional use. Cold cream in particular elicited a great deal of anal behavior. Two case histories are given comparing Rorschachs and the usage of these "sensory toys." The authors also used the techniques listed below but not for longitudinal study; experiments in group play and readiness for destruction; experiments in active play techniques (frustration and gratification techniques, marionette techniques, motivation techniques, and general diagnostic play techniques); observation of behavior in the nursery school which included records of routines and child's behavior in routine situations (rests, bathroom routine, etc.) in anecdotal or word-for-word format; summary of opinions of physical ability and creative activities; and observation during various situations including I.Q. tests and pediatric examinations.

163

LEVI, J., and KRAEMER, D. Significance of a Preponderance of Human Movement Responses on the Rorschach in Children below Age Ten, *Rorschach Research Exchange and Journal of Projective Techniques,* 16:361–365 (1952)

Setting: Clinic

Subjects: All children with an I.Q. over 90 seen in the clinic

Time span: 3 years

Meth. of obs. and test.: Case histories; Rorschach Test

Findings

Grouping of children — The children were divided according to the number of human movement responses. While 40 produced a normal number of M (average .8), 5 children showed abnormal increases (average 6).

Normal group — The I.Q. ran between 90 and 122. The ratio of FM:M gave higher FM, the type of M was mixed, and the ratio of M to the sum of color responses was greater to, equal to, or less than M.

Abnormal group (M's) — The I.Q. ran from 108 to 120, the ratio of FM:M gave lower FM, and M was always greater than the sum of color responses. The children tended to be passive, blocked, and depressed, and were lacking in play interest. All had some kind of school problem and difficulty in getting along with other children. All but one showed provocative, attention-getting behavior and temper tantrums, and all but one had experienced some sexual trauma, that is, they had seen either the primal scene or had a seductive parent. The parents of these children sought help primarily because of external pressure and not because of personal concern. One or both parents showed marked rejection and the mothers were markedly overprotective in all cases.

Authors' interpretations

The authors infer that a child needed all the above factors to become an M child since the normal group showed one or more but never all of these factors. They postulate that the M children were limited in their physical activity (overprotected) and were also forced to be more mature than was natural for them (parents' rejection of childish behavior). The marked aggression evident in these children was not handled constructively (rejection, punitive superego, introverted aggression). In times of stress they regressed to infantile patterns (tantrums, provocativeness, etc.) to release energies damned up by the repression of instinctual drives (M:FM). Movement may have been a defense aiding sexual repression. Some M responses were obviously wish-fulfilling.

164

Levy, D. M. *Studies in Sibling Rivalry* (Research Monographs, no. 2). American Orthopsychiatric Association, New York, 1937

Setting: Clinic

Subjects: 10 children, age 5 to 13 years; 12 children, age 3 to 4 years

Time span: Variable

Meth. of obs. and test.: Observations of directed play with a doll family; mothers' reports

Each case is outlined with patterns of hostility and complex reactions described and interpreted. The therapeutic aspects were studied through follow-up records which are appended.

Findings

Patterns differed from child to child but repeated themselves for each child in gradient or cyclical forms. Hostility, of which a rough measure is made, was manifested from slight movements to primitive crushing, tearing, and biting. Self-punitive behavior was seen in all but four of the 3- and 4-year-olds and appeared only with second and third degree (strong) hostility. "Restitution," or restoring of objects after they have been attacked, was noted. Every child used "defenses" (concealment, denial, projection) against consequences of hostility in proportion to frequency and intensity of hostility.

Author's interpretations

Therapeutic results are attributed to a reaction or acting out of the emotional drives in conjunction with interpretation through the medium of a parent substitute who does not criticize but who presents a friendly objective attitude. Follow-ups indicate a more favorable therapeutic change when there is a gradual positive acceleration to discharge of affect.

165

LEVY, J. A Quantitative Study of Behavior Problems in Relation to Family Constellation, *American Journal of Psychiatry,* 10:637–654 (1931)

Setting: Clinic and school

Subjects: 700 children referred to clinic because of personality or behavior problems (Group I); problem children from a wealthy community (Group II)

Time span: Not stated

Meth. of obs. and test.: Statistical analysis of case records; teachers' ratings of problem children

Findings

Community and family size as factors in behavior problems — In a clinical survey of the Chicago population, it was found that the distribution of children's behavior problems appeared to have little relation to family size. In a small rich community, families with an only male child may produce more problem children than other size family groups. In a large city boys were brought to a psychiatrist's attention twice as often as girls and the proportion was still higher in rich districts. Also in a large city the first-born was a problem child more frequently than children in another ordinal position, allowance being made for the fact that there *are* more first children; in the small high-grade community the second-born was the more frequent behavior deviant.

Sex of sibling as factor in behavior problems — The sex of siblings nearest in age to the problem child could have influence on the incidence of problem children. For the male problem child the nearest sibling above was male in 48 cases and female in 26 cases; the nearest sibling below was male in 41 cases and female in 36 cases. For the female problem child the nearest sibling above was male in 14 cases and female in 19 cases; the nearest sibling below was male in 24 cases and female in 12 cases.

Only children as factor in behavior problems — Only children had fewer personality and emotional difficulties but more scholastic difficulties than children from families having two children. Only children committed more acts of delinquency than a child with a brother or sister, but most delinquents came from large families where economic and social factors seemed more important than family interrelationships.

166

LEWIN, H. S. Changes of Attitudes Subsequent to Camp Experience, *Nervous Child*, 6:173–177 (1947)

Setting: 2 private camps — Camp A was traditional and adult-planned and Camp B was progressive and camper-planned

Subjects: 38 boys in Camp A; 46 boys in Camp B; age 12 and 13 years
Time span: Full camp session
Meth. of obs. and test.: Self rating scales; case histories

Findings

The boys in Camp B had greater changes in attitudes and maintained these changes after 3 months better than those in Camp A. The histories indicated that some of the attitudes changed by camp experiences were not transient and became integrated in the child's personality despite the absence of the original stimulus (camp).

Author's interpretations

The author points out that changes in attitude may result merely from maturation, vacation, and camp experience, but the fact that the boys from Camp B consistently had greater changes leads to the assumption that at least part of the change was due to the specific character and educational approach of the camp. Attitudinal changes following camp are not necessarily desirable; this depends on many factors such as overall camp morale, social controls, group pressures, and the type of guidance and leadership.

167

LEZINE, I. Recherches sur les étapes de la prise de conscience de soi chez les jeunes jumeaux, *Enfance,* 4:35–49 (1951)
Setting: Research institute
Subjects: 4 sets of identical male twins; nonidentical twins — 11 boys, 6 girls, 3 mixed
Time span: Not stated
Meth. of obs. and test.: Observations of free play; developmental tests; "Zazzo questionnaire"

Findings

Psychomotor development — The twins and a group of non-twins were compared for psychomotor development. The following chart shows the twins' relative slowness in development.

	Developmental quotients	
	Non-twins	*Twins*
Locomotion	73%	70%
Equilibrium	66	60
Play with cubes	70	66
Puzzles	66	60
Vocabulary	68	65
Comprehension	60	42
Self-sufficiency	68	42
Language	68	50
Social relations	66	42

Thus it was demonstrated that in the developmental quotients twins had a definite language deficit and difficulty in distinguishing between self and other. There were significant differences in the ability of the twins and non-twins to name themselves and parts of their bodies. Much of the twins' play was involved with the problem of interdistinction.

Twins' relations to objects — The development of a twin's relations to objects is traced in the following outline which illustrates certain peculiarities of the development of this faculty by twins. The author points out that the infant begins to manipulate objects as he becomes aware that they are part of a world separated from himself.

Age 1 to 6 Months
Body: Sucks thumb; plays with hands; catches feet
Objects: Plays with blanket
Provoked behavior: Sensitive to restraint
Mirror: Smiles
Twin: Cries if twin cries
Language: Vocalizes for the first time

Age 7 Months
Body: Puts feet in mouth
Objects: Changes and moves actively
Provoked behavior: Begins cooing; plays "here it is" with a towel; takes objects from head

Mirror: Touches image
Twin: Looks a great deal at twin
Language: Vocalizes and echoes

Age 8 Months
Body: Plays and crawls; moves in space
Objects: Finds; grasps; plays peek-a-boo; plays with materials
Provoked behavior: Lifts objects placed directly in hand
Mirror: Looks at image; inspects body parts
Twin: Looks a great deal at twin
Language: Vocalizes and echoes

Age 9 Months
Body: Stands
Objects: Uses objects
Provoked behavior: Lifts objects impeding feet and hands, but not those impeding back or chest
Mirror: Actively explores mirror; holds out play things to baby in glass
Twin: Explores body of other twin; looks for twin in mirror
Language: Vocalizes and echoes

Age 10 to 12 Months
Body: Begins to walk; feet change from play toys and become organs of activity
Objects: Uses varied objects; plays at exchanging
Provoked behavior: Begins body orientation; finds objects hidden on body (not only on extremities)
Mirror: Embraces image and makes social overtures
Twin: Momentary imitation of twin
Language: Speaks first words

Age 15 Months
Body: Walks
Objects: Plays at building
Provoked behavior: Plays hide-and-seek
Mirror: Acts sociably
Twin: Relates with twin by chance quarrels
Language: Babbles and once in a while use a pronoun common to the pair

Age 18 Months

Body: Shows parts of body on himself or twin

Objects: Uses complex material but plays alone

Twin: Shows confusion in play and quarrels over possessions

Language: Expresses presence and absence; associates words; orients himself verbally in space

Age 21 Months

Body: Points out body parts on doll

Objects: Imitative play

Twin: Shows organized imitation and decided ascendancy

Language: Confuses himself with twin and babbles to twin

168

LIPPITT, R. Popularity among Preschool Children, *Child Development,* 12:305–332 (1941)

Setting: Nursery school

Subjects: 45 children

Time span: Variable

Meth. of obs. and test.: Paired comparison method using ratings from peers and teachers; observations during 8 10-minute play periods; Ackerman's test of constructive and destructive behavior; Stanford-Binet Intelligence Scale; Keister's test; parts of McCaskill and Jakway tests; 30 1-minute observations for presence or absence of an oral nervous habit

Findings

Judgment of popularity — Adults and children judged popularity on apparently different criteria. Adults judged on behavioral criteria, i.e., teachers judged the socially active child popular, the isolated one unpopular. Children's preferences were apparently based on different criteria and did not agree closely with teachers' ratings. Popularity tended to remain constant.

Factors in popularity — In general there was a tendency for children to favor their own sex whereas teachers favored boys. If any one factor was indicative of child popularity, it was "cooperation in routines." It seems that quick adaptation to the situation without making a disturbance was typical of the most popular child while the reverse was true for unpopular ones.

Items inconsistently related to measures of popularity — These items included gross motor ability (related to teacher estimates), attractiveness (physical and personality), and the amount of hostile and friendly behavior.

169

LITHAUER, D. B., and KLINEBERG, O. A Study of the Variation in I.Q. of a Group of Dependent Children in Institution and Foster Homes (Chapter IX, Sections 1 and 2), in: *Twenty-seventh Yearbook: Nature and Nurture* (National Society for the Study of Education). Public School Publishing Co., Bloomington, Ill., 1928

Setting: Orphan asylum
Subjects: 120 children
Time span: Variable
Meth. of obs. and test.: Stanford-Binet Intelligence Scale

Findings

Dependent children placed in foster homes had a rise in I.Q. with M of 5.9.

170

LOOMIS, A. M. *A Technique for Observing the Social Behavior of Nursery School Children* (Child Development Monographs. no. 5). Teachers College, Columbia University, New York, 1931

Setting: Nursery schools
Subjects: 55 children; age 24 to 51 months
Time span: Variable from 3 months to 1 year
Meth. of obs. and test.: 15-minute observations noting especially physical contacts; measures of muscle tension

Findings

The total number of physical contacts was related to age;

physical contacts were also a fairly stable individual characteristic related to physiological tension and possibly language. All but 4 of 31 children increased their number of contacts during the study. Children with the highest muscular tension tended to avoid physical contacts but not necessarily social contacts. Significant individual differences were noted in the extent to which language, physical contact, or both were used for communication with associates.

Author's interpretations

It is the author's impression that at such an early stage in life a large proportion of the total social interaction is expressed by physical contact. Therefore studying physical contact has special relevance and indicates characteristic differences. It can be used as empirical methodology.

171

Lord, E. E. A Study of the Mental Development of Children with Lesion in the Central Nervous System, *Genetic Psychology Monographs*, 7:371–486 (1930)

Setting: Clinic

Subjects: 51 children with lesions in the central nervous system

Time span: Variable

Meth. of obs. and test.: Developmental tests; ratings of child and maternal attitudes; psychological case studies

Findings

The development of motor efficiency bore no direct relation to intellectual appreciation. Each child had a minimum response which could be related to a developmental sequence in him. Defective children consistently remained defective, but alterations in adjustment were evident between examinations. Development of mentality could not be predicted by the locus of the lesion. No child was inaccessible when the examiner was patient.

172

Lowenstein, P., and Svendsen, M. Experimental Modification

of the Behavior of a Selected Group of Shy and Withdrawn Children, *American Journal of Orthopsychiatry,* 8:639–653 (1938)

Setting: Camp and clinic

Subjects: 13 children characterized as shy; age 6 to 8 years

Time span: 6 months

Meth. of obs. and test.: Social work case histories; daily observations at camp

Findings

Improvement in behavior during the 8 weeks of camp — At first the children played alone, imitated each other, and strove for adult attention. Gradually they became more aggressive and interactive. Boy-girl play with marked sex interest appeared in the fifth week — i.e., pairing off, displaying affection, and acting out sexual interests. By the eighth week a marked increase in group play had developed. Even the very aggressive tended to be quieter following individual play periods.

Later improvement in behavior — Of the 9 who returned to the same environment, 5 had better family relationships, better school performance, and fewer neurotic symptoms after 4 months; 3 relapsed. All 4 children who returned to different environments showed better adjustment. There was improvement in 5 out of 9 eating problems, 6 out of 6 enuresis problems, 3 out of 4 disturbed sleeping problems, 3 out of 6 nervous mannerisms.

Authors' interpretations

Analysis of the data shows behavior changes were due chiefly to removal from tension-producing situations to settings where children interacted with positive, permissive adults. The authors conclude that even in 8 weeks some modification of behavior can be effected.

173

Lowrey, L. G. Personality Distortion and Early Institutional Care, *American Journal of Orthopsychiatry*, 10:576–586 (1940)

Setting: Orphan asylum

Subjects: 28 children
Time span: Variable
Meth. of obs. and test.: Case records

Personalities that are unsocial, isolated, and affectively distorted to a marked degree are found more often among children who are reared from infancy in institutions than among those who are reared in their own homes. The former undergo an isolation type of experience and react in essentially the same way as children who have been rejected. Children placed in institutions for short periods after the age of 2 years do not develop isolated personalities to the same degree or show the same behavior patterns.

174

Lurie, L. A., Levy, S., and Rosenthal, F. M. The Defective Delinquent; a Definition and Prognosis, *American Journal of Orthopsychiatry*, 14:95–103 (1944)
Setting: Guidance home
Subjects: 50 defective delinquents, age 3 to 20 years; 2 control groups — 25 nondefective delinquents and 25 defective nondelinquents
Time span: Variable
Meth. of obs. and test.: Physical and psychiatric examinations; psychological tests; personality ratings; interviews; general observations

Findings

Socio-psychological factors — Poor living conditions, broken homes, low morals, nonunderstanding parents and the like were characteristic of the backgrounds of the nondefective delinquents but not of the backgrounds of the defective nondelinquents

Family histories	*Defective delinquents*	*Nondefective delinquents*	*Defective nondelinquents*
Parent or sibling mentally retarded	80%	—%	32%
Parent or sibling with psychiatric conditions	72	84	48
Physical illnesses	62	52	40

Influence of gangs — Gang involvement ran as high as 64% for the nondefective delinquents and 32% for the defective delinquents, but only 4% for the defective nondelinquents.

Work history — The records showed that 44% of the nondefective delinquents, 28% of the defective delinquents, and 12% of the defective nondelinquents were disturbed in their work.

Personality traits — Generally both defective and nondefective delinquents had violent tempers and were suspicious, phlegmatic, egocentric, depressed, selfish, obstinate, intellectually dishonest, and emotionally unstable. In contrast, the defective nondelinquents were placid, trusting, emotionally stable, generous, etc. However, the two defective groups were similar in that they were largely imitative and lacked leadership qualities.

Authors' interpretations

The personality of a defective delinquent is arrested in the volitional, emotional, sexual, and intellectual areas. He constitutes a real clinical entity with vicious home influences and poor mental heredity. His personality is characterized as nonconforming, stubborn, resentful, insecure, and difficult to control in an institution. He feels wronged, is aggressive toward the world, and is unable to behave constructively. The child is blindly loyal to his family but strong sibling rivalry is usually apparent. He is physically and emotionally unattractive and mentally inaccessible; as a consequence, psychotherapy and social therapy are largely limited and ineffectual.

175

LURIE, L. A., LEVY, S., ROSENTHAL, F. M., and LURIE, O. B. Environmental Influences; the Relative Importance of Specific Exogenous Factors in Producing Behavior and Personality Disorders in Children, *American Journal of Orthopsychiatry*, 13:150–161 (1943)

Setting: Guidance home

Subjects: 400 children; age 3 to 18 years

Time span: Not stated

Meth. of obs. and test.: Not stated

Findings

Nearly all the personality disorders were due to the influence of the family; less than 1% could be directly traced to the neighborhood. However, gang influences were prominent in the formation of antisocial behavior. The more important disturbances were immoral conditions (a ratio of 9:1 for immorality), lack of understanding, and alcoholism in the parents. There was no relation be-between disorders and the economic status of the home. Most disturbances were due to a multiplicity of factors, but 80% of these factors were psychic. There were two unnatural homes for every natural one. Natural homes producing problem children were apt to have many physical and psychiatric problems.

176

Lurie, L. A., and Rosenthal, F. M. Military Adjustment of Former Problem Boys, *American Journal of Orthopsychiatry*, 14:400–405 (1944)

Setting: Guidance home

Subjects: 116 problem boys; age 5 to 20 years

Time span: Variable

Meth. of obs. and test.: Case histories; service reports

Findings

Adjustment prior to military service — Of the active group, 73 had made good social adjustment between discharge from the guidance home and induction in the Army, 42 were still maladjusted at the time of induction, and 1 could not be followed.

Military adjustment — Military adjustment was surprisingly good at the time of follow-up: 105 were still in military service, 7 were honorably discharged, 3 were dishonorably discharged, and 1 was killed. Promotions, special assignments, or decorations were given to 40% of the group.

Authors' interpretations

The authors feel that social maladjustment or neuropsychiatric disorders are not necessarily incompatible with normal adjustment to military life. None of the boys would have been accepted had the military authorities known their case histories.

177

Lurie, L. A., Schlan, L., and Freiberg, M. A Critical Analysis of the Progress of 55 Feeble-minded Children over a Period of Eight Years, *American Journal of Orthopsychiatry*, 2:58–69 (1932)

Setting: Hospital

Subjects: 55 children

Time span: 8 years

Meth. of obs. and test.: Physical examinations; psychiatric examinations; social histories; home visits; clinic records; social evaluations

Once diagnosed, these children were given social and vocational guidance, together with medical and occupational therapy.

Findings

Results of study — In 38% of the cases, recommendations were completely carried out and 95% of the subjects made successful adjustments. Where recommendations were partially carried out, most made only a partial adjustment; where recommendations were not carried out, 40% made no adjustment. Of the entire group, 60% made complete social adjustment, 23.6% made partial adjustment, and 16.4% made no readjustment. Twice as many girls as boys made successful adjustment and girls also showed greater stability than boys in their jobs.

Factors influencing adjustment — Good home influence was a distinct factor in permitting social adjustment, as was good health. Of the 33 who adjusted perfectly, only one was in poor physical condition whereas one-third of those not adjusted were in poor health. The lower the I.Q. of the mentally retarded child, the better his adjustment.

Authors' interpretations

The majority of feeble-minded children are potential assets to society and can become real assets with the proper therapeutic measures.

178

Mahler, M. S., Luke, J. A., and Daltroff, W. Clinical and

Follow-up Study of the Tic Syndrome in Children, *American Journal of Orthopsychiatry*, 15:631–647 (1945)
Setting: Hospital
Subjects: 16 male and 2 female patients
Time span: 2 to 15 years
Meth. of obs. and test.: Psychiatric case histories; psychotherapy

Findings

Onset of tic syndrome — The tic syndrome in its highly irreversible form does not crystallize in children before the latency period; before this there exists a behavior disorder in the form of either diffuse hyperactivity, dyskinesia, or impulsion, with seriously increased muscular tension and affectomotor overreactiveness.

Religious-cultural extraction — The racial pattern seems to facilitate the development of psychomotor neuroses. Among the 18 follow-up cases, 12 were Jewish, 5 non-Jewish, and 1 was of Catholic-Jewish origin.

Characteristics of children and parents — Oral overindulgence was one of the commonest features of the mother-child relationship. Only 3 children were thumb-suckers, 5 were enuretic; 60–70% either skipped or only very slightly indulged in creeping and suddenly developed upright locomotion. Motor development generally was not skillful or graceful. There was fear, sluggishness, or exaggerated recklessness; tension was indicated by restlessness, phobias, and temper tantrums. Children did not fulfill their potentialities for either academic or athletic achievement, although most had a desire to excel in the latter. The parents were perfectionistic and overprotective.

Psychosexual development — The psychosexual development of these children was partially fixated in attachment to the mother during the pre-Oedipal period. The tic syndrome is an equivalent of masturbation in the beginning at least.

Authors' interpretations

Follow-up data suggest that the prognosis for a child with a tic depends on many factors.

179

Mahler, M. S., Ross, J. R., and DeFries, Z. Clinical Studies in Benign and Malignant Cases of Childhood Psychosis (Schizophrenia-like), *American Journal of Orthopsychiatry,* 19:295–305 (1949)

Setting: Hospital

Subjects: 16 psychotic children

Time span: Variable

Meth. of obs. and test.: Psychotherapy and direct observation

The authors discuss and illustrate the pathognomonic clinical symptoms of psychoses in children.

Findings

Dynamic trends in psychotic development — There are three dynamic trends in psychotic development. The first (Group I) is an arrest of the differentiation and higher organization of the ego system resulting in a severe ego deficiency in the area of social development. The second (Group II) is a secondary regressive disintegration of an already formed but constitutionally defective and incohesive ego system. In the third (Group III) there is progressive severing of ties with reality up to age 10; this trend is characterized by the use of neurotic defenses.

Follow-up — The follow-up on Group I showed some personality integration and a low level of restitution. Group II showed greater deterioration and a proneness to panic and catatonia. Group III had a more benign course because larger parts of the personality remained intact.

Authors' interpretations

The psychotic process acts upon the personality of the child whose hold on reality is not consolidated. The differential diagnosis, evaluation of signs, and prognosis are all extremely difficult.

180

Mason, S. H. A Comparative Study of Four Pairs of Twins Examined in Kindergarten and in Junior High School, with Special Reference to Personality, *Smith College Studies in Social Work,* 4:197–286 (1934)

Setting: Clinic
Subjects: 4 pairs of normal twins
Time span: 7 years
Meth. of obs. and test.: Social case histories; ratings by parents and teachers; physical examinations; palm and sole configurations; psychometric tests; observations in controlled and uncontrolled situations; developmental tests; Woodworth-Cady Questionnaire; Stanford-Binet Intelligence Scale; achievement test

Full case descriptions of the four pairs of twins are presented.

Findings

Of the three pairs of monozygotic twins, only pair II remained similar during the 7-year observation period (1925–1932). Pairs I and III developed striking personality differences. Pair IV, the dizygotic twins, differed in appearance and personality from the beginning.

Author's interpretations

The striking differences between the twins of pair III may be accounted for by a serious illness suffered by one of the twins. The differences between the twins of pair I may have been related to factors in the emotional relationship of the twins and their parents. In both pairs I and III an apparent discrepancy in intelligence was found which may have been another factor in their personality. As none of the pairs was really "identical," it was suggested that the term be discarded.

181

McCaulley, S. One Hundred Non-conformed Boys, *Psychological Clinic*, 16:141–166 (1925)
Setting: 3 disciplinary centers
Subjects: 100 adolescent boys removed from school for disciplinary reasons
Time span: Variable
Meth. of obs. and test.: Social and scholastic histories; Revised Stanford-Binet Intelligence Scale; Witmer Formboard and Cylinder Test; Healy Construction Test; Dearborn Form-

board; proficiency tests in reading and arithmetic; auditory and visual-memory span tests; Woodworth-Wells Directions Test

Case examples are included.

Findings

Characteristics — The I.Q. range was 57 to 117 with only one score lower than 70. The I.Q.'s, however, must be weighed with many other factors — emotional, social, and educational. Most had poor language ability. The group lacked "that sort of mental control which correlated highly with intellectual ability," but the majority were socially well-oriented.

Test results — No single test showed nonconformity; the battery of tests proved helpful in understanding these boys, though there was no single trend. Every problem boy was an individual case. Truancy and schoolroom defiance were the principal offenses.

182

McCay, J. B., and Fowler, M. B. Some Sex Differences Observed in a Group of Nursery School Children, *Child Development*, 12:75–79 (1941)

Setting: Nursery school

Subjects: 31 girls and 35 boys enrolled in nursery school; median age, 2 years 7 months when school began

Time span: 5 years

Meth. of obs. and test.: Case histories; physical examinations; timed observations; observations of habit training

Findings

Boys showed higher food intake and more restless movement; they rose earlier and slept less per 24 hours. The authors discuss these points in terms of the greater size of the male and his tendency to show more aggressive extravert behavior than the female. The sex differences in the types of behavior studied were small and not statistically significant. No sex differences in growth rates or amount of sickness were found in these age groups during the observation period.

183

McCay, J. B., Waring, E. B., and Bull, H. D. Health and Development of a Group of Nursery School Children, *Child Development*, 11:127–141 (1940)

Setting: Nursery school

Subjects: 66 healthy children

Time span: 5 years

Meth. of obs. and test.: Records of behavior items coded and put on machines

Behavior was recorded on the following items: sleeping, eating, nervous behavior, bowel movements, enuresis, and outdoor play. Data were divided statistically into medians, interquartiles, and total range. The study presents normative values on these functions.

184

McCay, J. B., Waring, E. B., and Kruse, P. J. Learning by Children at Noon-Meal in a Nursery School; Ten Good Eaters and Ten Poor Eaters, *Genetic Psychology Monographs,* 22: 493–555 (1940)

Setting: Not stated

Subjects: 20 children; 10 good eaters, 10 poor eaters

Time span: 1 year

Meth. of obs. and test.: Records of eating behavior

An attempt was made to teach children to take what they wanted, to try all the foods, and to undergo a profitable learning experience in relation to food. Measurements were recorded on the amount of time spent in eating, the amount of food consumed, and efficiency in eating. The authors contrasted two groups.

Findings

Both groups, good and poor eaters, showed improvement over a period of a year. This was accomplished for example by offering a refused food frequently and in small amounts until it was accepted. Good eaters had slightly higher Merrill-Palmer scores, were better developed, and weighed more, but also had more colds.

Poor eaters had more G.I. disturbances and infections. The authors pointed out that there were no differences in breast feeding between the groups.

Authors' interpretations

The authors point out that eating is a frequent problem in children and that philosophies of feeding are confused. They consider mealtime a learning experience.

185

McFarlane, J. W. Some Findings from a Ten-Year Guidance Research Program, *Progressive Education,* 15:529–535 (1938)

Setting: Institute of Child Welfare, University of California (Berkeley)

Subjects: 252 children and their families; 126 guided and 126 unguided controls

Time span: 10 years

Meth. of obs. and test.: Perinatal data; systematic cumulative records of health, habit training, and growth; physical examinations; anthropometric measurements; developmental x-rays; photographs; records of mental development; records of personality development through observations of parents, siblings, teachers, classmates, clinic staff, and child himself; observations on environment; statistical analyses

DATA ANALYSIS

The analysis of data in the study was complicated by the variety of techniques used in obtaining data, the extensiveness of items used, and the long time span.

Findings

Analyses made in several areas — Analyses were made within four areas: (1) quantification of clinical material; (2) assessment of the reliability of material for statistical use; (3) group findings consisting of distributions and interrelationships over time, group trends, and changing interrelationships; and (4) individual findings consisting of the individual's trends as compared with group trends and with respect to himself.

Analysis of relationships between data from different areas — This analysis was done in terms of (1) cross-sectional interrelationships between single measures and between composites; (2) trends in relationships through time; and (3) "clusters" of relationships including biological findings, behavior manifestations, and environmental pressures.

SAMPLE FINDINGS

Problem behavior during preschool years — During preschool years no normal child was free of adjusting devices that could be called problem behavior. The average number of such problems was between 4 and 6; the frequency of most problems varied with age.

Physical condition — Many children showed defensive patterns to exceptional size and overreacted to irritating physical conditions. The oversized girl and undersized boy had greater problems of adjustment than those who fell in the middle area of size distribution. During preschool years, physical measurements had more prognostic value than mental measures. Elimination, eating difficulties, nail-biting, overdependence, and negativism all were more common among children of poor physiological status; thumb-sucking, overactivity, lying, and fears were more common among children with better-than-average physiological status. Temper and jealousy were found in children of both good and bad physiological status and seemed to be related more to psychological than physiological factors.

Intelligence — The prognostic value of mental test scores was poor for preschool years but increased in value with age. There were marked individual variations. Relationships of intelligence to personality and behavior problems were much smaller than those found in the case of either physiological condition or family influences.

Family variables — Correlations between the incidence of problems and factors in the home (socio-economic conditions, education, family constellations, and interpersonal relationships) were not high. Marital adjustment yielded higher correlations with behavior difficulties (demands for attention, tantrums, negativism, food finickiness, overdependence, and daytime enuresis). Thumb-sucking had no correlation with other problems and was taken from the favorable end of the scale on nearly all family variables. Parents' agreement on discipline showed a fairly high correlation with marital adjustment and much the same correlation with individual problems as obtained for marital adjustment. Fewer problems were found among children of relaxed mothers than among those of worrisome, tense mothers. Lower correlations were obtained with education than with intrafamilial adjustments. It was concluded that a child could usually fare well if only one or two psychological problems existed in the home.

Classroom reputations—Reputation among classmates showed marked individual differences in social approval, disapproval, and notice.

186

McFarlane, J. W. *Studies in Child Guidance; I—Methodology of Data Collection and Organization* (Society for Research in Child Development, Monograph, vol. 3, no. 6). National Research Council, Washington, D. C., 1938

Setting: Institute of Child Welfare, University of California (Berkeley)

Subjects: 252 children and their families; 126 guided and 126 unguided controls

Time span: 10 years

Meth. of obs. and test.: Perinatal data; systematic cumulative records of health, habit training, and growth; physical examinations; anthropometric measurements; developmental x-rays; photographs; records of mental development; records of personality development through observations of parents, siblings, teachers, classmates, clinic staff, and child himself; observations on environment; statistical analyses

This monograph describes the objectives and techniques of a 10-year study made by the Institute of Child Welfare of the University of California at Berkeley. Some background data on the subjects (as well as the actual test forms, questionnaires, and sample interviews) are included. Although the results of the study are not presented in this monograph, some corollary discoveries are discussed. For example, it was found that 9 out of 10 parents reported that the clinic had helped them solve the preschool problems of their children. Preliminary analysis of teachers' judgments of children then in early grades showed little difference between the guided and unguided groups except for a tendency in the former to show better verbal facility and social adjustment. Finally, the study served as a check on the cross-sectional findings and enabled the experimenters to judge better the prognostic significance of cross-sectional studies.

187

McFarlane, J. W. The Relation of Environmental Pressures to the Development of the Child's Personality and Habit Patterning, *Journal of Pediatrics*, 15:142–154 (1939)

Setting: Institute of Child Welfare, University of California (Berkeley)

Subjects: 252 children and their families; 126 guided and 126 unguided controls

Time span: 10 years

Meth. of obs. and test.: Perinatal data; systematic cumulative records of health, habit training, and growth; physical examinations; anthropometric measurements; developmental x-rays; photographs; records of mental development; records of personality development through observations of parents, siblings, teachers, classmates, clinic staff, and child himself; observations on environment; statistical analyses

These findings are based on a 10-year study made at the Institute of Child Welfare of the University of California at Berkeley. Most of the findings are discussed elsewhere; this paper deals mainly with environmental pressures.

Findings

In any area where parents were emotionally conditioned to anxiety by past experience or a present situation, they were apt to be unwise, exploitable, or uncertain. Children ironically chose the areas of the parents' greatest concern in which to develop disturbing habits. The marital adjustment of the parents had much more bearing upon the children's behavior, at least in preschool years, than any other factor in the home. Closely related was parental agreement or disagreement on discipline. Parent-child adjustment was another vital aspect for consideration.

188

McGraw, M. B. The Effect of Specific Training upon Behavior Development during the First Two Years, *Psychological Bulletin*, 31:748–749 (1934)

Setting: Child development clinic

Subjects: 1 pair of twins; 1 trained and 1 untrained

Time span: First 24 months of life

Meth. of obs. and test.: Developmental tests; Thematic Apperception Test

Findings

"The great differences in the performances of the two children and the greater differences in their emotional adjustments to the particular experimental situations were remarkably reduced by the two and a half months training of the control twin at the age of twenty-two months. . . . There is no evidence that the early training of the experimental baby was of great advantage to him later in the solution of distinctly new problems or in making a rating on intelligence tests."

189

McGraw, M. B. *Growth; a Study of Johnny and Jimmy*. Appleton-Century, New York, 1935

Setting: Laboratory and home

Subjects: 1 pair of twins; 1 trained and 1 untrained

Time span: 26 months

Meth. of obs. and test.: Developmental tests; Thematic Apperception Test; motion pictures

Arrangements for the study were made prior to birth and therefore motion pictures of the delivery and the first 15 minutes of life were obtained. Johnny was delivered after a breech presentation, Jimmy after a vertex presentation. Since Jimmy was physically dominant and Johnny flaccid by comparison, Johnny was picked for the training program. The twins were kept at the laboratory from 9:00 to 5:00 o'clock 5 days a week, Jimmy being placed in a crib during these training periods. The training continued until the twins were 26 months old at which time they went home and had bi-weekly follow-ups.

Findings

Training development — Johnny was startled 40 times a week to elicit the Moro reflex; this reflex was not affected by practice, nor were suspension grasp, inverted suspension, creeping, first steps, or sitting. Jimmy stood up earlier, but Johnny did it more maturely later. Around 7 months of age training differences became more marked. The trained twin began swimming at 231 days, began to swim without support at 308 days, and could do it alone and enjoy it at the age of 1 year at which time he had also learned to dive. The control twin could not perform any of these activities. In other complicated activities such as roller skating, climbing, and jumping the trained twin excelled.

Timing of training — The author emphasized that the timing of training was crucial — training if not correlated with maturation can cause tension and may develop into fear because of growth imbalances. As the trained twin was more persevering than the control in new situations, it was inferred that he was not particularly frightened. After a month's abstinence the trained twin's skills deteriorated in all performances while the controlled twin's skills remained constant.

Author's interpretations

The author concludes that behavior patterns begin with random body development and then vague and uncontrolled development of certain movements. Phases of the complete behavior pattern develop later and may complement or interfere with each other. Finally all phases mature, providing for integral action. It is noted that behavior tends to regress with the increasing difficulty of a task. Certain traits are subject to little if any modification through practice; other performances can be greatly improved through exercise. The effect of this training on subsequent development is the subject of future study.

190

McGraw, M. B. Later Development of Children Specially Trained during Infancy; Johnny and Jimmy at School Age, *Child Development*, 10:1–19 (1939)

Setting: Not stated

Subjects: 1 pair of twins

Time span: 6 years

Meth. of obs. and test.: Developmental tests; Thematic Apperception Test

This is a report on Johnny (T) and Jimmy (C) who as twins were treated experimentally. Johnny was given special training in motor activity; Jimmy was not.

Findings

Follow-up tests of a pair of twins indicated that Johnny (the trained twin) had better coordination than Jimmy (the untrained twin) but that Johnny's lead in motor skills depended on the fixation of the skill at the end of the training period. The more fixed the skill at that time, the more marked was Johnny's superiority at follow-up. Psychological tests revealed that Johnny was enuretic, had a richer fantasy life including the wish to grow up to be a giant, had played with fire, etc. This child, who had been trained from a very early age in various motor activities when 6 years old showed many traits of the Icarian personality described by H. A. Murray; Jimmy's personality was much less complex.

191

McGregor, H. G. Enuresis in Children; a Report of Seventy Cases, *British Medical Journal*, 1:1061–1063 (1937)

Setting: Hospital, school, and private practice

Subjects: 70 enuretic children
Time span: 7 years
Meth. of obs. and test.: Detailed micturition history of patient and family; physical examinations; children's records of their own successes and failures

Findings

Family records — Of the total, 7% of the children had at least one parent with a history of enuresis and 14% had a sibling with a similar record. As for family size, 75% came from families having 3 or more children, 10 from families of 2 children, and 9 were only children. The ratio of hospital ward patients to private patients was 2.3:1.

Characteristics — Associated conditions included threadworm (35%), enlarged tonsils and adenoids (21%), undescended testicles (3%), and petit mal epilepsy (1.4%). The children were divided into three temperamental groups — nervous or excitable (15%), normal (70%), and phlegmatic (15%).

Technique and results of treatment — Each child was given a bedside calendar on which he recorded his dry nights. Reward, praise, and encouragement made up the treatment. Results showed that 60% maintained the cure from 18 months to 6 years, 10% were "cured" for only a few months, 10% were not followed-up, and 20% failed.

Author's interpretations

"Most of the children [can be persuaded] to change the habit by simple encouragement or other suggestive means; nor is the task very laborious."

192

McKinnon, K. M. *Consistency and Change in Behavior Manifestations* (Child Development Monographs, no. 30). Teachers College, Columbia University, New York, 1942
Setting: Nursery school
Subjects: 16 children
Time span: 5 years for 2 subjects; 6 years for 14 subjects

Meth. of obs. and test.: Written evaluations of behavior put into itemized forms evolving four categories — conformity, caution, invasiveness, and withdrawal; 12 semiannual teachers' reports; achievement ratings; physical and intelligence records; interviews with parents and teachers

Findings

Findings at age 3 to 4 years — At this level, 5 showed conformity predominantly, 5 invasiveness, 3 withdrawal, and 3 caution.

Follow-up at age 8 to 9 years — Of the same 16 children, 10 continued to show the dominant characteristics they had at age 3–4. Withdrawal was most consistent; all 3 continued to show this trait from age 3–4. Of the 3 who showed caution, 2 persisted and 1 shifted to conformity. A conformist shifted to caution. Invasiveness persisted but became less frequent as a predominant characteristic (5 at age 3, and 3 at age 8), while conformity increased. When there was a change in behavior it tended to be more in the direction of conformity than in other directions.

Adult judgments — These revealed that the most wholesome form of behavior according to adults was that labeled as conformity in this study; but conformity may be associated with characteristics that tend to limit personal development. Invasive behavior at age 3 was considered necessary and acceptable by adults, but at age 8 was disapproved by peers as well as adults.

Behavior as influence on adjustment — Predominant invasiveness tended to interfere with personal-social development. Early predominance of caution was found to be detrimental to the development of a well-rounded personality. However, the predominantly cautious children all adjusted very well to adult-directed activity and routine aspects of school life. There was some indication that improvement in specific skills may have a beneficial influence on adjustment, but the relation was not a consistent one. Inability to attain a favorable social adjustment seemed to motivate achievement in school work in some children, particularly at the older age levels. At the 8- or 9-year level, 9 children had noticeable difficulty in working harmoniously with their peers as a result of predominant caution, withdrawal, or invasiveness. Persisting predominant modes of behavior sometimes became less conspicuous with age. In cases where the educational philosophy at home differed from that at school, the children made less satisfactory adjustment to school life in all phases.

Health as factor of invasiveness — The data suggest that predominantly invasive behavior may be influenced by health factors since 4 invasive children at age 3 were underweight, malnourished, and readily fatigued.

Innate traits as factor in behavior — The findings suggest that innate traits alone were not responsible for the predominant forms of behavior. Innate tendencies perhaps influenced the intensity or direction of forms of behavior motivated by environmental factors. Innate intellectual abilities seemed to foster cautious behavior in the case of 2 children, but 2 others with superior I.Q.'s showed either invasiveness or conformity.

Author's interpretations

Modification of behavior is a gradual process of development. Change appears to be largely a matter of degree rather than of kind.

193

MECHEM, E. Affectivity and Growth in Children, *Child Development*, 14:91–115 (1943)

Setting: Elementary school

Subjects: 30 boys and 35 girls who were in a growth study

Time span: 1 year

Meth. of obs. and test.: Interviews including 71 specific questions; I.Q. tests; Haggerty-Olson-Wickinan Problem Tendency Test

Findings

Factors influencing affectivity — Children with high academic grades had high positive affectivity. On retesting, individual childdren showed Pearson correlations of .45 between increasing growth and increasing affectivity. Children with overall high "organismic" age showed high affectivity (R = .43); those with gains in I.Q.

showed gains in affectivity and vice versa (R = .28). Many other relationships of less statistical significance were reported for the group.

Report on extreme of affectivity — There was an additional report on four children representing highest and lowest affectivity, and greatest gain and loss in affectivity. The child who had the highest affectivity was small in size and scored high in achievement. This child had a low Haggerty-Olson-Wickinan problem tendency score. The one who had the lowest affectivity was large in size and scored low in achievement and high in the problem tendency score. The child with accelerated growth had the greatest gain in affectivity and went down in the Haggerty-Olson-Wickinan score. The child with a plateau in development had the greatest loss in affectivity; this child increased his Haggerty-Olson-Wickinan problem tendency score.

Author's interpretations

The author concludes that affectivity is related to overall growth.

194

Melcher, R. T. Development within the First Two Years of Infants Prematurely Born, *Child Development*, 8:1–14 (1937)

Setting: Psychological institute

Subjects: 42 premature infants

Time span: Not stated

Meth. of obs. and test.: Bühler-Hetzer Infant Test; observations during free play; interviews with parents; hospital records

Findings

Comparison of premature with full-term babies — Quantitative test analyses showed that premature children lagged behind the averages of full-term babies up to the age of 5 months but scored within the average limits thereafter. Qualitatively, retardation of up to 18 months was noted in postural control. Predominant personality traits in premature babies were positive reactions, dependence upon social stimulus and response, and moderate affective reactions. They were characterized as "gentle" babies.

Birth weight as factor in development — There was a low positive correlation between birth weight and developmental quotients. Peculiarities of appearance were more likely to persist into the second year when the birth weight was under 2,000 grams than when it was over 2,000 grams.

195

MEYERS, M. R., and CUSHING, H. M. Types and Incidence of Behavior Problems in Relation to Cultural Background, *American Journal of Orthopsychiatry*, 6:110–116 (1936)

Setting: Clinic

Subjects: 3 groups of American-born children with symptoms of maladjustment — 50 of primarily British descent, 50 of south Italian descent, 50 of Russian-Jewish or Turkish-Jewish descent

Time span: 2 years

Meth. of obs. and test.: Case records; interviews

Findings

Problem categories — There was an average of 6.3 problems per child. The categories were socially unacceptable behavior (truancy, sex problems, criminality); faulty habits and defects (enuresis, speech or health problems, poor work, smoking, nervousness, and poor adjustment); adverse family relationships; and personality difficulties (instability, seclusiveness, feelings of inferiority, neurosis, fear, egocentricity, and hyperactivity).

National and racial differences — Jewish children had the most faulty habits and defects, adverse family relationships, and personality difficulties. Children of Italian descent had the most socially unacceptable behavior (stealing, lying, sex misconduct, running away, and untidiness) and poorer health. Children of British descent were lowest in all four categories. Despite differences in behavior, differences in personality were quite small for the three groups.

Types of symptoms and possible influences — The authors

pointed out that there were obvious differences in the types of symptoms displayed by different groups — e.g., overt behavior differed markedly, and parental harshness or neglect had higher incidence among the children of Italian descent perhaps because of the low educational opportunities of parents and low economic status. Problems among Jewish children were quite different. They suffered from feeding difficulties, parental ignorance, maternal overprotection and immaturity, and seclusiveness. This may be accounted for by a narrow emotional and intellectual heritage leading to overintensification of familial affectional ties. Children of British descent were divergent in having excessive nervous habits, coming from broken homes, and being unwanted children.

196

MILNER, E. Effects of Sex Role and Social Status on the Early Adolescent Personality, *Genetic Psychology Monographs*, 40: 231–325 (1949)

Setting: Home and school

Subjects: 149 children, with extensive and intensive study of 15 boys and 15 girls

Time span: 4 years; from 10 to 14 years of age

Meth. of obs. and test.: Repeated interviews; Stanford-Binet Intelligence Scale; Primary Mental Abilities Test; Otis Mental Ability Test; Wechsler-Bellevue Intelligence Scale; Cornell-Coxe Performance Ability Scale; Iowa Silent Reading Test; Minnesota Paper Formboard; Chicago Assembly; Minnesota Mechanical Assembly; Porteus Maze; Metropolitan Achievement Test; Stanford Achievement Test; California Personality Test; school grades; family relations questionnaire; Chicago Interest Inventory; strength of conscience questionnaire; emotional response test; Bavelas Moral Ideology Test; essays; reputation ratings; Thematic Apperception Test; Rorschach Test; behavior ratings by teachers; sociometric data; anthropometric data; clinical conferences summarizing the above data

Findings

Personality traits — Of the 39 traits worked out for analysis, 15 were typical of girls, 13 were typical of boys, and 11 were typical of both sexes. In the absence of affection, adolescents tended to experience a high level of anxiety. This led to a reduction in mental efficiency. Active conformity was used to obtain more love as a basis for personal adjustment.

Sex role and social status — Home training and mother-child relationships were found to be responsible for deviation and typicality in the group under observation. Since parents had a clearer idea of what was right for girls than of what was right for boys, the girls were more preoccupied than boys with social conformity and suppression of impulses. There was a strong resemblance between the personalities of the girls in the group and that of the typical housewife, suggesting that major sex-delineated characteristics are developed quite early. At age 10, boys were still preoccupied with the control of impulses whereas girls had already suppressed them. It was harder for girls than for boys to relate positively and spontaneously with others. Major sex differences in personality were established by adolescence.

Author's interpretations

Children's personalities reflect the parents' adherence to the standards and methods of the group as well as social status per se. The author emphasizes the importance of training agents in a given subculture. Normal and neurotic forms of behavior may be conceptualized in terms of the interaction of individual genetic forces with culturally derived and culturally deviant forces. The author compares normal and neurotic behavior in these terms.

197

MONTALTO, F. Maternal Behavior and Child Personality, *Journal of Projective Techniques*, 16:151–178 (1952)

Setting: Home and research institute

Subjects: 57 children; age 6 or 7 years

Time span: 38 months

Meth. of obs. and test.: Fels Parent Rating Scale; Rorschach Test with Monroe Checklist items

A case study is presented.

Findings

Cluster analysis — Maternal behavior patterns were determined by a "cluster analysis" of the Fels Parent Rating Scale. The clusters contained scales of warmth, democracy, intellectuality, restrictiveness, and indulgence. There were four clusters used: (I) high in democracy and low in the other categories; (II) high in restrictiveness and low in other categories; (III) high in warmth and democracy, average in intellectuality and indulgence, and low in restrictiveness; (IV) high in indulgence, warmth, and restrictiveness, and low in democracy.

Intergroup differences — The Rorschach scores were transposed into Monroe checklist items which were analyzed with exactness to determine the extent of intergroup differences. They revealed a correlation between maternal attitudes and children's personality characteristics. Of 7 intergroup differences, 6 referred to Group II (5 were emotional, 1 emotional and intellectual).

Experience types — "Affectively stable" characterized children whose records showed a balance of movement and color not greater than 2.1 (M was equal to, or less than, 4). "Affectively unstable" characterized children whose records showed color predominance. "Coarctated" characterized children whose records showed flat affect.

	Groups			
Experience type	*I*	*II*	*III*	*IV*
Coarctated	2	1	4	1
Affectively stable	6	4	10	6
Affectively unstable	7	8	2	16
Total	15	13	16	23

Analysis of intergroup differences — Exact analysis showed a pattern of emotional unstability for Group II and a pattern of emotional stability for Group III. This supported the differences found in color dimensions among the groups, particularly in Group II. The two democratic groups (I and III) showed no common trends. Differences in ratings of warmth, indulgence, and intellectuality did have selective influence on Rorschach trends, but high democracy and low restraint did not. Restrictive groups (II and IV) showed common trends on CF and color shock. The child was reactive to outer emotional influences (IV) and was less secure in emotional contacts (II). Groups I and IV showed pressure for achievement (high W± and F%). In Groups II and III, W ± was low and W form high.

198

Mooney, M., and Witmer, H. L. Problem Children Who Later Became Psychotic, *Smith College Studies in Social Work,* 2:109–150 (1932)

Setting: Clinic and hospital

Subjects: 10 problem children; age 10 to 19 years

Time span: Variable

Meth. of obs. and test.: Social work case histories

The 10 patients' case histories are briefly presented. All were between 10 and 19 years of age when first seen at the institute; 7 were 15 years of age or younger. In no case was the possibility of psychosis among the reasons for requesting examination.

Findings

The examining psychiatrists recognized symptoms prognostic of a psychosis in only 3 children, 3 were described as psychopathic, 1 was "psychoneurotic," 1 was given a good prognosis, and 2 were not diagnosed. All the patients were seclusive, had few or no friends, and violent tempers.

199

Moore, E. S. *The Development of Mental Health in a Group of Young Children; an Analysis of Factors in Purposeful Activity* (Studies in Child Welfare, Monograph, vol. 4, no. 6). University of Iowa, Iowa City, 1931

Setting: Preschool nursery and home

Subjects: Approximately 20 children; supplementary studies of more extensive numbers

Time span: 1 year

Meth. of obs. and test.: Comprehensive observations, including observation in a controlled experimental situation twice a year; rating scales; home visits; home histories from parents

Comprehensive observations were made at school and at home; then generic terms classifying and categorizing the data were chosen. This is called by the author the "inductive method." An attempt was made to record teacher-child relations systematically. The study focused on initiative, creative ability, perseverance, self-reliance, and friendliness. The children were rated in terms of their most marked deviation in behavior and adjustment and were then compared with children deviating in the opposite extreme.

Findings

Characteristic behavior — After comparing the extremes of deviation it was found that there was striking individuality in each child's behavior and also a modifiability of behavior. Thus the children became less deviant with repeated measurements; this was attributed to opportunities for free play.

Changes in first and second semesters — There was a change from one semester to the next in the types of teacher-child interaction wherein the teacher assisted the child. In order of frequency, the first semester children sought companionship, did not adjust to social needs, explored, sought affection, and showed instability. During the second semester (also in order of frequency) they again sought companionship, now enjoyed books and pictures, conversed with others, explored, and showed self-assertion.

Significance of home behavior — "Parent firmness" tended to correlate with self-reliance; "wholesome play life" with self-assertion; "cheerfully matter-of-fact" with continued effort; "stressful mealtimes" with seeking-demonstrative. No correlation was found between economic adequacy and child stability.

Summary — Wide individual differences existed as early as age 2. Home environment and choice of play material varied greatly; nonetheless, nearly all deviant behavior moved closer to the norm during the second semester. Over the year the author noted that teachers tended to react to different children in strikingly different ways, but they reacted to any given child in the same way.

200

Morgan, S. S., and Morgan, J. J. An Examination of the Development of Certain Adaptive Behavior Patterns in Infants, *Journal of Pediatrics,* 25:168–177 (1944)

Setting: Institution

Subjects: 18 infants

Time span: Observations made at frequent intervals from birth to age 90 days

Meth. of obs. and test.: Records of visual and social responses

The 5,300 observations procured were reduced to a percentage basis, and smoothed curves were constructed for each of the activities recorded. A developmental scale was devised in which critical points extended from the first 10 days of life to the 71–75 day period.

201

Mowrer, O. H., and Mowrer, W. M. Enuresis; a Method for Its Study and Treatment, *American Journal of Orthopsychiatry,* 8:436–459 (1938)

Setting: Clinic

Subjects: 30 children; age 3 to 13 years

Time span: Varied

Meth. of obs. and test.: Case histories

The authors devised a therapeutic method for the solution of enuresis by providing a means of awakening the child right after the onset of urination, thus associating the need to urinate with the act of awakening. The child slept on a pad which when wet caused a bell to ring which woke the child who then went to the toilet to finish. All other therapy was discontinued while this method was being tested and this practice was continued until the child had gone seven consecutive nights without wetting his bed.

Findings

All 30 children, including a feeble-minded child who responded satisfactorily, were cured of enuresis in a maximum time of 2 months. Personality changes, if any, were uniformly favorable and no "symptom substitution" was evident.

Authors' interpretations

A survey of the literature indicates to the authors that most writers believe enuresis (1) is a continuation of infantile incontinence due to poor training, and (2) is caused by emotional needs which do not find expression in waking hours — e.g., genital sexuality, deep-rooted anxieties, or hostility to parents. In the authors' opinion these latter emotional factors are predominant only in isolated cases. In a large proportion of enuretic children faulty training is the only significant factor. The therapeutic method devised by the authors is not advised for very young, mentally underdeveloped, or psychotic children.

202

MUMMERY, D. V. An Analytical Study of Ascendant Behavior of Preschool Children, *Child Development*, 18:40–81 (1947)

Setting: Nursery school

Subjects: 42 children; age 3 and 4 years

Time span: 6 weeks

Meth. of obs. and test.: Diary records in the "Jack experimental situation"

Findings

The seven least ascendant children showed a statistically significant increase in ascendant behavior after training.

203

MURPHY, L. B. Sympathetic Behavior in Young Children, *Journal of Experimental Education,* 5:79–90 (1936)

Setting: Nursery school

Subjects: Not stated

Time span: Variable

Meth. of obs. and test.: Observations and records of specific episodes; teachers' ratings; experimental situations; interviews with parents; parents' day-to-day records

The author has compiled classification of the stimuli and responses characteristic of sympathy.

Findings

Responses — In one-half of the instances where a given child could react to another in a distress situation, both sympathetic and unsympathetic responses were given on different occasions. In one-third of the cases a child responded consistently with sympathetic responses; in one-sixth of the cases he responded only with unsympathetic responses to a given child. Sympathetic behavior correlated with intelligence, aggressive behavior, and cooperativeness. No sex differences existed.

Stimuli — In a group with a wide age range (30 months) there were 26% more sympathetic responses than in a group with a narrow age range (10 months). Other factors influencing behavior were teachers' response patterns to sympathy situations, effects of behavior of dominant personalities in the group, and the effect of the number of children in relation to the space.

Author's interpretations

On the average, one sympathetic response can be expected per hour of observation of a child, or one-eighth as many sympathetic responses as conflicts for the same group. Sympathetic responses which ordinarily appear in certain situations are inhibited when the ego of the potentially responsive child is threatened.

204

MURPHY, L. B. *Social Behavior and Child Personality.* Columbia University, New York, 1937

Setting: Nursery school

Subjects: 2 groups; 20 in one, 19 in the other

Time span: 8 months' intensive observation for 2 hours per day

Meth. of obs. and test.: Teachers' ratings; interviews with parents; experimental situations to test children's reactions to artificially introduced stimuli

Findings

Of two groups studied, one showed a higher percentage of sympathetic behavior than the other.

Author's interpretations

This evidence of more sympathy may be due to a more overtly sympathetic teacher, more space for play and therefore fewer conflicts, or a wider range of age which fostered sympathetic responses from the older children toward the younger ones. Sympathy is to be considered not as an entity but as one facet of the total personality. It is related to age, sex, intelligence, background, culture, the group, and the influence of specific situations in and out of school. The ways of showing sympathy are also related to these variables. The author also points out that sympathy requires a secure ego and that 3-year-olds who are afraid to cry cannot understand why people cry.

205

MURPHY, L. B. Personality Development of a Boy from Age Two to Seven, *American Journal of Orthopsychiatry,* 14:10–16 (1944)

Setting: Not stated

Subject: 1 boy

Time span: 5 years

Meth. of obs. and test.: Behavior records; observations of play; Rorschach and other projective tests; Lerner's Ego Blocking Test; medical examinations

A life history is presented which is composed of teachers' records and notes, observations during free play and play with miniature dolls, standard projective tests, and other experimental observations. The data were set down with the Rorschach data over a similar time and a blind analysis was made of the latter.

Findings

The child had an unusual capacity for affective behavior but the Rorschach test at that time showed an independent ego with a potential for great aggression and a hint of anxiety. This tendency became manifest in his subsequent behavior. At other points also the Rorschach hinted at behavior which later became manifest.

Author's interpretations

The opinion was offered that this child had a verbal talent which permitted him to express subsequently what might have remained at the fantasy level in other children. Other general conclusions were derived from the study. One was that clinical entities such as fear of bodily mutilation, anxiety, extreme aggression, etc. can appear during the course of development of a normal child. Also a struggle against this fear may result in the release of creativeness; the ebb and flow of fear and fight creates a superficial variability in behavior, but seen longitudinally the striking continuity of self is evident. The extent to which success in achieving growth can conquer anxiety may depend on the child's satisfaction and confidence in his ability to manipulate the environment. Concluding, the author suggests that while behavior varies in its manifest form, the inner core remains stable. Therefore one cannot judge personality from behavior at a particular age except by relating behavior to the external situation and its meaning to the child.

206

NEILON, P. Shirley's Babies after Fifteen Years, *Pedagogical Seminary and Journal of Genetic Psychology,* 73:175–186 (1948)

Setting: Not stated

Subjects: 15 of original 19 babies who had been in Shirley's longitudinal study for 2 years

Time span: 15 years

Meth. of obs. and test.: Goodenough Speed of Association Test; Minnesota Survey of Opinions Test; trait scales and special abilities; standardized interviews with children and mothers; fathers' ratings of children by mail

Findings

Judges were asked to compare the character sketches of the

2-year-olds and the 17-year-olds. The sketches based on the above tests were matched correctly in a statistically significant manner.

Author's interpretations

The study confirms Shirley's original hypothesis that significant personality traits persist.

207

NELSON, V. L., and RICHARDS, T. W. Studies in Mental Development; I — Performance on Gesell Items at Six Months and Its Predictive Value for Performance on Mental Tests at Two and Three Years, *Pedagogical Seminary and Journal of Genetic Psychology*, 52:303–325 (1938)

Setting: Not stated

Subjects: 31 children

Time span: 3 years

Meth. of obs. and test.: Gesell Developmental Schedule; Merrill-Palmer Scale; Revised Stanford-Binet Intelligence Scale

Findings

Items connected with manipulation correlated low in relative significance to later I.Q. tests, whereas those items connected with perception of objects at a distance and social awareness had greater relative significance.

208

NELSON, V. L., and RICHARDS, T. W. Studies in Mental Development; III — Performance of Twelve-Months-Old Children on the Gesell Schedule, and Its Predictive Value for Mental Status at Two and Three Years, *Pedagogical Seminary and Journal of Genetic Psychology*, 54:181–191 (1939)

Setting: Not stated

Subjects: 123 children; age 6 months

Time span: 30 months

Meth. of obs. and test.: Gesell Developmental Schedule; Merrill-Palmer Scale; Revised Stanford-Binet Intelligence Scale

Findings

Reaching, grasping, and manipulation constituted the central core of the Gesell test at 6 months; distance perception and awareness were less important at this age for Gesell but became more significantly related to later tests of mental development. The Gesell at 6 months showed a correlation of .37 with the Merrill-Palmer scale at 2 years and a correlation of .46 with the Stanford-Binet scale (a poor correlation).

209

NEWBERY, H. Studies in Fetal Behavior; IV — The Measurement of Three Types of Fetal Activity, *Journal of Comparative Psychology,* 32:521–530 (1941)

Setting: Research laboratory

Subjects: 30 pregnant women

Time span: Last 2 to 5 lunar months of pregnancy

Meth. of obs. and test.: Modified polygraph records triggered by mothers

The data of this experiment are contrasted with the theories of Coghill who claims that mass action develops before individual reflexes, and those of Windle who claims that anoxia increases as the fetus develops and that increased squirming results. Three types of fetal movement are described — squirming, kicking, and hiccuping.

Findings

From 6 to 8 months kicking is the dominant fetal movement; after 8 months squirming is the more frequent. Hiccuping increases slightly with the later development of the fetus. Correlations of retest scores revealed reliable and significant differences between fetuses in the amount and type of activity.

210

NEWELL, H. W. The Psycho-Dynamics of Maternal Rejection,

American Journal of Orthopsychiatry, 4:387–401 (1934)
Setting: School
Subjects: 33 unwanted children; age 5 to 18 years; median age, 11 years
Time span: Variable
Meth. of obs. and test.: Social work case histories; psychiatric examinations

Findings

Causative factors — Maternal rejection was directly related to four factors. Of the cases of rejections studied, 36% could be attributed to a sacrifice of social life because of childbirth, 27% to fear that the child would inherit bad tendencies, 18% to disgust or fear concerning pregnancy, and 15% to the fact that the marriage was forced by pregnancy. Rejection was indirectly due to unhappy marital adjustment usually the result of immaturity or emotional instability on the part of one or both parents.

Expression of rejection — Of the mothers, 59% expressed their rejection of the child by neglect and cruelty, 22% by overprotection, and 18% by inconsistency. Of the children, 54% openly expressed the feeling that their mothers lacked affection for them. Children, being particularly sensitive to attention and the need to feel welcome and important, derived much satisfaction from having their mothers upset about them. Much of their difficult behavior represented the discovery of what their mothers feared the most. When the mother's handling was mixed, the child's behavior usually was nonaggressive or mixed. There was a clear trend of interrelationships — e.g., the more overt the parent's rejection, the more frequent the child's aggressive behavior.

211

Newell, H. W. A Further Study of Maternal Rejection, *American Journal of Orthopsychiatry,* 6:576–589 (1936)
Setting: Clinic
Subjects: 75 unwanted children (age 4 to 18 years; median age, 10 years); 82 controls in grades 3 and 5 (median age, 9 years)
Time span: Variable

Meth. of obs. and test.: Psychiatric case histories; psychiatric interviews with children. Controls had in addition school records, physical examinations, teachers' reports, and classroom observations by psychiatric social worker.

Data is included in tables with summaries.

Findings

The children showed a mixture of aggressive, antisocial, and submissive-neurotic symptoms. Aggressive behavior was more frequent when the parent's handling was consistently hostile, while submissive behavior was more frequent when the parent's handling was consistently protective. The mother's handling wavered between overprotective and hostile behavior and was most frequently inconsistent. The most important cause for rejection was the mother's unhappy adjustment to marriage. This maladjustment to marriage was often due to immaturity or emotional instability on the part of one or both parents. The control group indicated the correlation between constructive parental handling and stable behavior on the part of the child.

212

NEWMAN, H. H., FREEMAN, F. N., and HOLZINGER, K. J. *Twins; a Study of Heredity and Environment.* University of Chicago, Chicago, 1937

Setting: Not stated

Subjects: 50 pairs of identical twins compared with 50 pairs of fraternal twins; 19 pairs of identical twins reared apart since infancy are studied intensively

Time span: 10 years

Meth. of obs. and test.: Anthropometric ratings; intelligence and educational achievement tests; Downey-Hill Temperament Test; Woodworth-Mathews test; data from parents and others on health, school history, and interest

Findings

Identical twins were found to be much more alike than fraternal twins with respect to all the traits studied, except those classified under personality. Differences in mental traits but not in physical traits increased with age.

Authors' interpretations

The authors suggest that the increasing differences between twins are due to the influence of environment — i.e., physical traits are affected least, intelligence somewhat more, scholastic achievement still more, and personality most. "It may be that coarser outlines of behavior, including perhaps those that are most susceptible to quantitative measurements, are relatively more determined by genetic constitution; whereas the finer details which may be observed but are difficult to measure are more subject to modification by the environment, or even by chance."

213

OBERNDORF, C. P., ORGEL, S. Z., and GOLDMAN, J. Observations and Results of Therapeusis of Problem Children in a Dependency Institution, *American Journal of Orthopsychiatry,* 6:538–552 (1936)

Setting: Institution

Subjects: 25 male and 25 female orphans

Time span: 1 year of treatment and 10 years of follow-up

Meth. of obs. and test.: Data from psychoanalytic psychotherapy; social work histories

Findings

Of the 50 children treated, 11 boys and 15 girls made a satisfactory life adjustment 5 years after graduation; 8 boys and 6 girls required occasional assistance; 4 boys and 3 girls encountered serious difficulty; and 2 boys and 1 girl were either dead or out of touch.

Authors' interpretations

Psychiatric aid and removal from home environment both play a role in the improvement of these children. Normal children in institutions do not seem adversely affected by problem children.

214

Olin, I. The Social Adjustment of Children of Low Intelligence; Follow-up Study of 26 Dull-Normal Problem Children, *Smith College Studies in Social Work,* 1:107–159 (1930)

Setting: Not stated

Subjects: 26 dull-normal problem children

Time span: 3 to 5 years after clinic treatment

Meth. of obs. and test.: Case records; psychological tests; records of psychiatric treatment; interviews with mothers; achievement and behavior ratings by teachers

The criteria selected for social adjustment included the child's situation in the home, symptomatic behavior problems, friends and interests, home duties, school progress, and classroom attitudes.

Findings

Poor adjustment was made by 77% of the girls but by only 41% of the boys. Health and physical defects were closely related to adjustment; beneficial changes in home and school environments were associated with increased adjustment. In general, no correlations were found between age, I.Q., economic status, and the original problems. During the follow-up period the original problems of sex misconduct, enuresis, and feeding disappeared, while truancy, temper sensitiveness, and untidiness continued.

215

Parten, M. Social Participation among Preschool Children, *Journal of Abnormal and Social Psychology,* 27:243–269 (1932)

Setting: Preschool nursery

Subjects: 42 preschool children; age 2 to 4 years

Time span: 8 months

Meth. of obs. and test.: Time-sampling observations; teachers' ratings; observations during free play

Findings

Frequency of types of participation — Unoccupied behavior was noted in 5 children, onlooking in all but 2, associative play in all but 1, and solitary play in all. During the observation period individual children spent from 1–57% of their time in cooperative play. Parallel activity, in which two-thirds of the children spent 33% of their time, was the most common form of behavior.

Correlation of factors with social participation — The dependence of social participation on age was shown by the correlation of .61. Despite much individual variation, younger children tended to play alone or in parallel groups, and older children in more highly organized groups. I.Q. and social participation correlated .26 whereas parallel play correlated with I.Q. .69. Nursery school experience correlated .12 with participation. Positive social participation increased over the observation period while negative participation decreased for the group as a whole.

216

PARTEN, M. Leadership among Preschool Children, *Journal of Abnormal and Social Psychology,* 27:430–440 (1933)

Setting: Preschool nursery

Subjects: 34 children

Time span: Not stated

Meth. of obs. and test.: Time-sampling observations of behavior; teacher's ratings

Findings

Behavioral changes related to age — Independent play is most characteristic of all ages but decreases in frequency as children grow older. Following, directing, and reciprocal directing increase with age. Individual differences in following, directing, and independent pursuit outweigh differences attributable to age.

Leadership characteristics — As the school year advances there is a trend toward development of leadership. Even at the age of 2 years definite types of leaders are apparent, e.g., the "diplomat" and "bully." The former artfully and indirectly controls a large group, while the latter bosses a small gang. Leaders somewhat exceed nonleaders in intelligence and generally come from a higher occupational class. Sex differences in leadership are negligible.

Author's interpretations

As leadership was consistently and reliably recorded by the method used, the author feels that the scores are probably valid since they agree with the teacher's ratings of the trait.

217

PAULSEN, A. A. Personality Development in the Middle Years of Childhood; a Ten-Year Longitudinal Study of Thirty Public School Children by Means of Rorschach Tests and Social Histories, *American Journal of Orthopsychiatry,* 24:336–350 (1954)

Setting: School

Subjects: 30 children

Time span: 10 years

Meth. of obs. and test.: I.Q. and Rorschach Test in first grade with retests every 2 years; social work case histories; school records; interviews with children and guidance personnel

Findings

Intellectual and personality development — Rorschach data indicates that the greatest changes in intellectual development take place from 6 to 8 years of age and there is a sharp rise in form accuracy and attainment during this period. From 8 to 10 years of age there is little intellectual change, but there is a steady rise in texture responses, which shows a new sensitivity to the environment. Also the greatest changes in color and movement responses take place and at 10 years of age the ratio of form to color exceeds color to form for the first time. The highest incidence of constriction in individual records occurs in this period. Disturbed content is prominent at some time in the records of all of the children as well as spurts, plateaus, and regressive movements. At 12 years of age there is a new spurt with increased attention to detail and heightened productivity.

Human and animal movement — There is an increase with age of both human and animal movement, animal movement exceeding human movement throughout. Early high movement occurred in two children, one from a deprived home and one from a home where the mother displayed a markedly neurotic dependency on the child.

Experience balance — The M/C ratio characterizing introversive and extratensive did not remain constant; 40% showed extreme shifts. This, it is felt, reflects the actual change in these children in relation to their contacts with internal and external realms of experience — e.g., whereas 17% showed introversive ratios at 6 years of age, 40% showed these ratios at 12 years of age; and whereas 47% showed extratensive ratios at 6 years of age, only 17% displayed them at 12 years of age.

Form responses — In general, the children had a higher form percentage than adults, e.g., 60–68% as compared with 50% in adults. This is paradoxical since F% according to the usual Rorschach measures describes the degree of conscious control. The two possible explanations given for this paradox are that it is an artifact due to fewer total responses and less elaborated responses, and that it may represent an expenditure of ego energy to attain stability which is not necessary in adults. It is important to note that children with low F% were from troubled homes.

Comparison with follow-up study — The several protocols of a child bore the stamp of his "individuality." Every child in the group showed positive achievement in terms of personality development. The follow-up study of 21 subjects with ratings made of adjustment level showed that 13 Rorschach tests agreed with the evaluative rating. A comparison of the evaluation between ratings at age 16 and the Rorschach test at age 6 showed agreement in 10 cases and wide disagreement in none.

Author's interpretations

The author believes that the development of personality is intimately associated with maturation and that this developmental picture is consistent with the Rorschach theory. A correspondence was found between Rorschach and case history material, but evaluative data and Rorschach Tests are best combined for interpreting the child's dynamics and prognosis. At 6 years of age 71% of the adjustment rating pattern was already established.

218

Pearson, G. H. Some Early Factors in the Formation of Personality, *American Journal of Orthopsychiatry,* 1:284–291 (1931)

Setting: Clinic

Subjects: 72 problem children under 7 years of age

Time span: Not stated

Meth. of obs. and test.: Not specifically stated

Findings

Influence of pregnancy on formation of personality — Children whose mothers were worried or unhappy during pregnancy showed traits which, if taken together, would constitute a maladjusted personality. These traits included fretfulness in infancy, physical weakness, late speech development, defective speech, masturbation, fearfulness, oral habits, enormous appetite, and maladjustment in school. Other characteristics included alertness and artistic ability. This picture correlated also with premature birth, rejective attitudes from one or both parents, and overindulgence (usually from the father).

Influence of birth duration on formation of personality — There was a difference in behavior between children whose birth time was short and those whose birth was long. Long-birth children tended to be somewhat more neurotic and to be eldest children, only children, or the children of oversolicitous parents (frequently with opposing attitudes toward child rearing).

Influence of nursing on formation of personality — Over half the problem children were nursed for less than 6 months compared with the average of 6 to 10 months. Infant placidity, solitary play, antagonism to siblings, disobedience, school maladjustment, timidity, sex play, and sulkiness were more frequent among children nursed less than 6 months or more than 10 months. The factors associated with nursing less than 6 months were illness of mother during pregnancy, premature birth, instrumental birth, father's antagonism to pregnancy, rejection on the part of both parents, and the discouragement of the mother and indulgence of the father. The factors related to nursing more than 10 months were the mother's discomfort, depression, and illness during pregnancy; mother's antagonism to pregnancy; father's indulgence, mother's oversolicitousness; and mother's impatience.

Influence of parental attitudes on formation of personality — A study of birth and nursing situations shows that parental attitudes exert a more important influence on the formation of the child's personality than the actual events.

Author's interpretations

Formulations imply that specific events do not fix the type of personality. The parental attitude is a symptom of parental adjustment or lack of it and the child comes to symbolize the total parental maladjustment. The essential element for success in therapy will be the adjustment of the parents rather than changes of specific child-rearing practices.

219

PETERSON, C., and SPANO, F. Breast-Feeding, Maternal Rejection, and Child Personality, *Character and Personality,* 10:62–66 (1941)

Setting: Home and nursery school

Subjects: 126 nursery school children

Time span: Approximately 3 years

Meth. of obs. and test.: Records of parent care; Fels Parent Rating Scale; Fels Child Behavior Scale; Joël Behavior Maturity Scale; Vineland Social Maturity Scale; Brown Personality Inventory

In this study of 126 cases, the maximum duration of breast-feeding was 104 weeks and the mean duration was 10.2 weeks (S.D. = 18.92 weeks). The maximum duration of unsupplemented breast-feeding was 65 weeks and the mean duration 10.01 weeks (S.D. = 13.29 weeks). All correlations between feeding and child personality were too low to be statistically significant. The authors point out that the direction of the correlations was often contrary to psychoanalytic theory. No correlation was found between duration of breast-feeding and maternal rejection.

220

Portenier, L. The Psychological Field as a Determinant of the Behavior and Attitudes of Preschool Children, *Pedagogical Seminary and Journal of Genetic Psychology,* 62:327–333 (1943)

Setting: Nursery school

Subjects: 30 children

Time span: 7 years overall; 9 months of organized observation

Meth. of obs. and test.: Time-sampling observations of nervous habits; records of routine habits; rating scales

Findings

There were slight differences noted in the ease of adjustment to the nursery school for low socio-economic groups, but there were no differences in nervous habits and emotional status.

Author's interpretations

The author concludes that the particular nature of the child and his relation to the total character of the psychological field in which he must function determines his behavior.

221

Proctor, S. M. Psychological Tests and Guidance of High School Pupils, *Journal of Educational Research,* 1:1–125 (1923)

Setting: High school

Subjects: 272 students

Time span: 4 years

Meth. of obs. and test.: Records of vocational choice and eventual outcome

At the end of 4 years 56% of the girls and 26% of the boys were in the same or higher ranking vocational groups (40% of the total). The author believes that vocational preferences are out of proportion to available opportunities in the community.

222

Prugh, D. G., Staub, E., Sands, H. H., Kirschbaum, R., and

Lenihan, E. A. A Study of the Emotional Reactions of Children and Families to Hospitalization and Illness, *American Journal of Orthopsychiatry*, 23:70–106 (1953)

Setting: Hospital

Subjects: 2 groups of 100 children matched in relation to age, sex, diagnosis, etc.

Time span: 6 months to 1 year

Meth. of obs. and test.: Medical histories; observations on the wards; psychiatric interviews with parents and children; independently kept records by psychiatrist, psychologist, play supervisor, head nurse, and social workers; developmental data on physical growth

A control group under study for 4 months was subjected to the traditional ward experience — weekly 2-hour visits from parents and the usual attention from personnel. An experimental group had daily visits from parents, early ambulation, play program with trained teacher, psychological preparation for, and support during, diagnostic and therapeutic procedures, etc.

Findings

Group Reactions

Immediate reactions — All children had some reactions to hospital experience. Immediate reactions were most marked from 2 to 5 years of age.

Factors influencing immediate reactions — The adjustive capacity of the individual child correlated highly with his immediate ward adjustment, but not with his later reactions; nor were there correlations between length of stay and reaction or adjustment. The successfully adjusted child generally had satisfying relations with his mother and well-adjusted parents. Neither stress, judged objectively, nor the type of illness (even of central nervous system) bore significant relationship to the degree of reaction in either group. Sex differences were not striking. No conclusions could be drawn concerning the length or nature of the psychological preparation for admission or the subsequent reaction.

Long-range reactions — Of the control group, 92% showed significant difficulties in adaptation, while only 68% of the experimental group had difficulties. After 3 months, 58% of the controls and 44% of the experimentals showed at least moderately disturbed reactions; 10% of the experimentals were considered better adjusted after discharge. In follow-up after discharge 41% of the controls and 45% of the experimentals who remained disturbed were under 4 years of age; 54% of the controls and 69% of the experimentals still disturbed were under 6 years of age. Six months after discharge not more than 15% of the controls showed continued disturbances, but the controls' disturbances remained more severe. Both the younger and the older children in the control group showed disturbances, whereas the younger children in the experimental group were affected much more often than the older children. There was also a high correlation between the appearance of crippling behavior disturbances on the ward and the persistence of these disturbances after discharge.

Individual Reactions

Types of reactions — Overt anxiety was the most common disturbance at all ages in both groups. A variety of defensive and restitutive mechanisms were employed which were related to the level of psychosexual development and personality structure prior to hospitalization. Reactions to specific procedures were dealt with according to the characteristic anxieties and fears of the particular level of libidinal development.

Age 2 to 4 years — Controls displayed fear or anger at parents' leaving, constant crying, panic with adults, somatic anxiety (urinary frequency, diarrhea, vomiting), depression, withdrawal, need of tangible evidence of home (doll, clothing, etc.), regression in feeding and toilet behavior, sleep disturbances, habit disturbances (sucking, rocking), restlessness and aggressive acting out, and denial of illness and loss. The experimentals showed the same disturbances but less frequently and less severely — e.g., aggressive behavior, thumb-sucking, and rocking occurred only half as often. However, more expression of guilt was noted in this group.

Age 4 to 6 years — At this age overt anxiety, depression, pregenital gratification, hyperactivity, training regression, and aggression were found less frequently; but phobia and somatization reactions were found more often. Acceptance of dependent needs was particularly hard for this group, especially the boys. The experimentals tended to identify with aggressors and manifested only half as much anxiety as the controls.

Age 6 to 8 years — The anxiety present in the control group during these years revealed fewer panics and was more "free floating" or attached to definite experiences. There were fewer pregenital needs and less somatization and aggression. But more compulsive, obsessive, restless behavior, many conversion symptoms, denial, and strong rationalization were noted. Fantasies involved mutilation and death, and illness was regarded as a magical retribution for unacceptable impulses or acts. The experimental group resorted to less strenuous defenses, depending rather on verbalization and acting out in play of their fantasies, hostility, and fear. The experimentals took advantage of the opportunities for sublimation and temporary superego identifications.

Post-Hospitalization Reactions

Regressive behavior in young children subsided soon; anxiety due to separation lasted a long time in young and latency age groups. Demanding, infantile, dependent behavior persisted for several months. The symptoms of the experimental group were milder and fewer at home. The control group displayed twice as much increased anxiety at the check-up.

Parent and Family Reactions

Parents reacted in various ways to the illness and hospitalization of their children. The observers noted manifestations of fear, anxiety, guilt, denial, and ambivalence. Increased sibling rivalry often developed in the home. Well-adjusted parents liked the experimental plan.

223

Pyle, M. K., and McFarlane, J. W. The Consistency of Reports on Developmental Data, *Psychological Bulletin*, 31:596 (1934)

Setting: Clinic

Subjects: Not stated
Time span: 21 months
Meth. of obs. and test.: Comparison of mothers' later reports with developmental records

Actual records of the subjects' development up to 21 months of age were compared with information later supplied by the mothers on the basis of recollection. The only factor on which there was high agreement was birth weight. Agreement was markedly low on the mother's condition during pregnancy, and only fair on the age at which the child walked alone. In general, the mother saw the child with "a halo of precocity."

224

Radke, M. J. *The Relation of Parental Authority to Children's Behavior and Attitudes* (Institute of Child Welfare, Monograph Series, no. 22). University of Minnesota, Minneapolis, 1946

Setting: Nursery school and kindergarten
Subjects: 43 children; age 3 years 10 months to 5 years 10 months
Time span: Variable
Meth. of obs. and test.: Questionnaires to fathers and mothers; interviews with children; special projective techniques; observations of doll play; special experimental situation to judge compliance; teachers' rating scales

Findings

Comparison of child-rearing disciplines given and received by parents — When the discipline received by the parent in his childhood was compared with that used by him in raising his own children, various changes were found in the latter. On the whole the disciplinary situation was less emotional and greater responsibility was assumed by the father. Automatic methods of control were used less often and there was better parent-child rapport.

Projective techniques and interviews — The use of projective techniques (dolls and pictures) was discussed. Projections of the home situation were indicated in the children's responses. Their answers in interviews were devoid of adult rationalization and were quite clear. Some relationships were found between the scores on home discipline derived from parents' inventories and the rating of children's behavior in preschool and experimental situations.

Children's concept of parents — Children described parents in the abstract as people engaged in the routines of daily work and living. They seldom defined the personalities of their parents or expressed affection for them. The mother was perceived as the more influential authority and there was some evidence of differential girl-parent and boy-parent relationships.

Children's concepts of behavior and discipline — Children's standards of good and bad emphasized behavior which would fit into the adult's routine and would avoid his displeasure. They reported spanking and isolation as the most frequent disciplinary measures but stated that the fear of punishment by parents was not a motivation for better behavior. In general they advocated as proper the discipline they were receiving.

225

RICHARDS, E. L. Dispensary Contacts with Delinquent Trends in Children; Group I — Forty-Eight Cases of Stealing, *Mental Hygiene*, 8:912–947 (1924)

Setting: Henry Phipps Psychiatric Dispensary

Subjects: 48 children classified as deliquents because of stealing; 13 mentally defective, 35 normal

Time span: 1 to 3 years

Meth. of obs. and test.: Binet-Simon Intelligence Test; developmental histories; medical examinations; interviews with children; agency reports

Findings

Results of follow-up study — Of the nondefective group, 29 made complete adjustment; 23 who made this adjustment were outside of correctional institutions and 6 adjusted to institutional life; 2 were diagnosed as psychopathic personalities and 1 died.

Categories of stealing — On the basis of the data from tests and examinations, seven categories of stealing were differentiated: stealing as a gratification of normal childish cravings for sweets, toys, bright colors, etc.; stealing as motivated by a spirit of adventure; stealing as a satisfacion of a need for money to be used as a tool in securing social equality; stealing as a result of bad companionship and inadequate opportunities for play and other expressive outlets; stealing as an expression of a specific emotional reaction; stealing as an outgrowth of parental dishonesty and inconsistency (this group posed the greatest problem in therapy); and stealing as a part of the psychopathic personality.

226

Richards, E. L. Dispensary Contacts with Delinquent Trends in Children; Group II — Abnormal Sex Trends, *Mental Hygiene*, 9:314–339 (1925)

Setting: Henry Phipps Psychiatric Dispensary

Subjects: 29 children with sex misconduct; age 4 to 16 years at first examination

Time span: Variable

Meth. of obs. and test.: Binet-Simon Intelligence Test; developmental histories; medical examinations; interviews with children; agency reports

A comparison was made between the 10 children described as having "defective constitutional equipment" and the other 19 children.

Findings

Types of sexual problems — The main types were promiscuity, perversion, and autoeroticism. These problems were complex in origin and content.

Factors in family background — The main type of sexual problems seemed to evolve from current environmental influences rather than from the child's instinctual cravings. Desertion, alcoholism, low standards (economic, social, and moral), mental defectiveness, and psychoses were evidenced in the family history. Parental training of children showed general lack of supervision, faulty methods of personal hygiene, and inconsistency.

Follow-up results — Of the 19 normal subjects, 14 adjusted adequately; of the 10 defectives, only 1 adjusted. Changes in environment brought improvement in a short time in certain cases; in only 4 cases were good results obtained when the child remained in the same environment.

227

RICHARDS, T. W. The Individual Child's Development as Reflected by the Rorschach Performance, *Rorschach Research Exchange and Journal of Genetic Psychology*, 12:59–64 (1948)

Setting: Fels Research Institute

Subjects: 2 children; 1 boy, 1 girl

Time span: Tested at 5, 6, 10, 11, 13, and 16 years of age

Meth. of obs. and test.: Rorschach Test

The author states that certain characteristics which might be considered basic from a psychoanalytic standpoint are shown by Rorschach protocols to be fluctuant in the dynamically changing picture of personality.

228

RICHARDS, T. W., NEWBERY, H., and FALLGATTER, R. Studies in Fetal Behavior; II—Activity of the Human Fetus *in utero* and Its Relation to Other Prenatal Conditions, Particularly the Mother's Basal Metabolic Rate, *Child Development*, 9:69–78 (1938)

Setting: Fels Research Institute

Subjects: 17 mothers

Time span: Last 20 weeks of pregnancy

Meth. of obs. and test.: Polygraph records of mothers' subjective experience of fetal movements; basal metabolic rates of mothers; heart rates of infants

Findings

Correlation of fetal activity and mother's metabolism — Fetal activity increased during pregnancy, then dropped off in the last month. A rising basal metabolism rate over a month of pregnancy correlated significantly with the amount of fetal activity (R = +.60).

Correlation of position to placement — The rank position of various fetuses revealed that some fetuses tended to be consistently more active during gestation.

Correlation of fetal activity and birth size — There was no direct relation between the size at birth and the activity of fetus; nor were the active fetuses necessarily the ones with high heart rates.

229

RICHARDS, T. W., and NEWBERY, H. Studies in Fetal Behavior; III — Can Performance on Test Items at Six Months Postnatally Be Predicted on the Basis of Fetal Activity? *Child Development*, 9:79–86 (1938)

Setting: Fels Research Institute

Subjects: 12 infants and mothers

Time span: Prenatal to 6 months and 1 year

Meth. of obs. and test.: Polygraph records of mothers' subjective experience of fetal movements; Gesell Developmental Schedule

Findings

There was a consistently positive correlation between fetal activity and scores on the Gesell development chart (R = .62 ± .18 at 12 months). There were four suggested explanations for this correlation. First, the mothers who tended to report more fetal activity may have endowed their offspring with greater ability genetically. Second, these mothers may have surrounded their children with environmental factors conducive to mental development. Third, the correlation between fetal movement and postnatal development may have been due to the facilitation of 6 months' ability by the exercise of function itself. And fourth, the amount of time during which a fetus is active may be a criterion of the developmental pattern or patterns of behavior responsible for rather specialized performances postnatally.

230

Richards-Nash, A. A. The Psychology of Superior Children, *Pedagogical Seminary,* 31:209–246 (1924)

Setting: School

Subjects: 70 children with I.Q.'s over 120

Time span: Variable

Meth. of obs. and test.: School records; teachers' reports; principal's reports; medical records; Stanford-Binet Intelligence Scale; Pintner Educational Survey Test; interviews with children, parents, and teachers; general observations

The article includes discussions of definition, classification, and achievements of so-called superior children, and presents biographies of famous men and women. Of the 70 children observed, 10 had literary ability, 3 drawing, 11 dramatic, 5 musical, and 5 scientific.

Findings

Types of social interaction — The three distinct types of successful social interaction found in this group were unobtrusive followers, individuals who maintained their own interests but nevertheless remained well-liked by their schoolmates, and five unmistakable leaders. There was also a significant problem group.

Causative factors in the problem group — The four causative factors discussed were physical disability, emotional disability, channeling of energy along lines of special nonacademic interests, and lack of motivation — i.e., work was too easy.

Characteristics of unstable children — High learning rate, hyperactivity, poor conformity to school routine, erratic moods, omniverous reading, no specific or permanent interests, and poorly integrated personalities were characteristics of unstable children.

Author's interpretations

The author emphasizes that traditional ideas of social adjustment are not applicable to superior children, who are "normal" in relation to their few similar friends. She also points out the inadequacies of mental tests as criteria for evaluation of superiority: there is no test which eliminates the spuriously superior — e.g., the prematurely developed; superior children often work below capacity; and individual differences in personality and achievement must be equally emphasized.

231

Ripin, R. A Comparative Study of the Development of Infants in an Institution with Those in Homes of Low Socio-economic Status, *Psychological Bulletin*, 30:680–681 (1933)

Setting: School

Subjects: 10 subjects and 10 controls in each of the following age groups: 4 to 6 months, 7 to 8 months, 9 to 10 months, 11 to 12 months; healthy babies in Home for Hebrew Infants and in low socio-economic homes

Time span: Variable

Meth. of obs. and test.: Gesell and Bühler developmental scales; observations of infant attitudes to persons and objects

Findings

No differences were apparent between the two groups among children less than 6 months old, but marked differences became apparent with increasing age; the private-home group fared better. Ranking from least to greatest in degree of contrast, the differences were postural, linguistic, mental (memory, imitation, etc.), social, manipulative, and relational items.

232

Roberts, K. E., and Ball, R. S. A Study of Personality in Young Children by Means of a Series of Rating Scales, *Pedagogical Seminary and Journal of Genetic Psychology,* 52:79–149 (1938)

Setting: Nursery school and clubs

Subjects: Not stated

Time span: Not stated

Meth. of obs. and test.: Rating scales

The central purpose of the paper is to demonstrate the use of rating scales in the study of personality development.

Findings

Repeated ratings over a period of years showed great variation and indicated fluctuation in personality development. Personality was measured by means of ratings in nine major categories: response to authority, respect for property rights, tendency to face reality, independence of adult affection and attention, attractiveness of personality, physical attractiveness, ascendance-submission, sociability, and compliance with routine.

233

ROBINSON, E. W., and CONRAD, H. S. The Reliability of Observations of Talkativeness and Social Contact among Nursery School Children by the "Short Time Sample" Technique, *Journal of Experimental Education*, 2:161–165 (1933)

Setting: Nursery school
Subjects: 50 children
Time span: 1 semester
Meth. of obs. and test.: Time sampling

Findings

The children's behavior varied considerably from time to time on different days, but definite correlations were found between scores on talkativeness and social contact and between age and scores on both factors. There was a split half reliability, i.e., a .95 coefficient of talking measurement and a .99 coefficient of social contact.

234

ROFF, M., and ROFF, L. An Analysis of the Variance of Conflict Behavior in Preschool Children, *Child Development*, 11:43–60 (1940)

Setting: W.P.A. nursery school
Subjects: 33 children
Time span: Not stated
Meth. of obs. and test.: Continuous 15-minute records of a diary type, with special note of conflicts

Findings

Characteristics of children in group — These children who belonged to a low socio-economic class, displayed less attention-getting behavior, less resistance to the teacher's suggestions, and more sharing of materials and play.

Comparison with Jersild group — The nursery school group showed much less conflict than did the group studied by Jersild. This difference was thought to be due to factors in home and school environment.

Authors' interpretations

The authors conclude that data obtained from observations of children's behavior depend on the group chosen. The type and amount of conflicts vary with different groups. Thus, if a problem child in this group was compared with one in Jersild's group, he would fall in the middle range as to number of conflicts.

235

Rogerson, B. C. F., and Rogerson, C. H. Feeding in Infancy and Subsequent Psychological Difficulties, *Journal of Mental Science*, 85:1163–1182 (1939)

Setting: Clinic

Subjects: 109 children

Time span: 6 months of observation and follow-up at 5 years

Meth. of obs. and test.: Medical records and examinations; home visits; school records

Findings

The evidence strongly suggests a positive association between feeding difficulties in infancy and psychological difficulties at school age.

236

Ross, B. Sibling Jealousy in Problem Children, *Smith College Studies in Social Work*, 1: 364–376 (1931)

Setting: Clinic

Subjects: 166 children said to have sibling jealousy compared with 1,109 children who lacked this trait
Time span: Not stated
Meth. of obs. and test.: Case records

Findings

It was found that there were more girls than boys, more bright children than dull, and on the average more young (most often under 6 years) than old in the jealous group. Small families had more jealousy, but this may be an artifact because small families refer more young children; in two-child families especially, more first-born were jealous. Maternal overprotection was equal in jealous and nonjealous groups; nagging mothers and those who made unfavorable comparisons were found relatively more often in the jealous group. Of the behavior problems studied only negativism and fears (to a lesser degree) were associated with jealousy.

237

SANDLER, R. J. (trans. Eugénie Evart-Chmielniski) Rapports réciproques entre les enfants au cours de la deuxième année de leur vie, *Enfance,* 6:1–47 (1953). Originally under: Bibliothèque des problèmes de développement et de l'éducation de jeunes enfants, no. 5, Éditions d'État de la littérature médicale, Moscou, 1950
Setting: Institution
Subjects: 9 children; 6 infants, 3 1-year-olds
Time span: 1 year
Meth. of obs. and test.: Systematic observations; uninterrupted observations for 24 hours 3 to 5 times a month on 4 of the children

NEGATIVE BEHAVIOR

Negative behavior is divided into two basic groups, offensive and defensive. The former encompasses all situations in which the child initiates the action; if it is aggressive, it is called aggressive-offensive behavior; if it is not aggressive, it is called active-offensive. Defensive behavior is the reaction of the child to intervention by another child. These reactions may be purely passive (for example, when the child merely cries) or they may be active and aggressive; these are called, respectively, passive-defensive and active-defensive behavior.

Findings

Active-offensive behavior — This behavior (taking away objects by force) begins when the child is between 8 and 10 months of age. It tends to wane in the second year and is characterized by the child's not using the object once he has it. At the beginning of the second year perseverative, almost reflexive examples of this behavior occur. Taking objects by force in the second year becomes linked with play activities, setting toys in order, or some favorite toys. The stimulus of an object exciting offensive behavior is greater when the object is held in another child's hand.

Active-defensive behavior — In the first year, the child already responds to the visual cue of another child approaching to take a toy. By the second year this set of defenses has evolved and become more complex; the complexity of the response is relative to the child's experience. Also there is an increasing exchange of objects and requests for desired objects; and in the second half of the year active handing over of requested objects begins. The author feels that this change in interaction is effected partly by the educative efforts of parents and teachers and partly by the child's personal experience in the community. She believes that children are more likely to give up toys when asked politely than when attacked. This sharing behavior is traced from the first to the third year.

Aggressive-offensive behavior — This behavior occurs when one child tries to seize an object from another and when he attacks another child who has begun to cry. In the latter instance the author attributes this response to a conditioned connection between tears and aggression. The observers noted but were unable to account for various other acts of aggression. Since aggression tends to become linked with all sorts of divergent behavior, it does not fit easily into categories. Aggressive quarreling and interchange appear when the child is about 20 months of age and sometime during the second half of the second year the abused child begins to turn to an adult for a solution.

Passive-defensive behavior — The stimulus for this form of behavior is the same as for active-defensive responses. At the beginning of the second year when an object is taken by force or activity is interfered with, the child may respond with passive behavior; at the end of the second year the child may simply let another child hit him or may weep without hitting back. Generally children at this age begin to weep when frustrated. However, holding grudges and other more involved feelings become noticeable and complicate passive defenses. Passive reaction like crying may be transformed into active techniques to get the teacher to intercede.

Indefinable behavior — The observers also noted some indifferent passive behavior which could not be categorized, such as surrendering an object when another child wanted it. Disputes of ownership which are neither aggressive nor passive sometimes occurred. These disputes quickly tended to become linked with adult attention and jealousy.

Changing negative behavior to positive — The author believes that prevention of negative behavior is implemented by an organized system of standards, adequate toys, and free play and that negative behavior can be transformed into positive behavior by experience and training. She also believes that language abilities and the role of the adult are important factors.

POSITIVE BEHAVIOR

Positive behavior cannot be arbitrarily separated from negative behavior. Attention is focused on intermediate forms, such as asking and exchanging, which become more frequent during the second year. Positive relations are divided into positive emotional behavior, common manipulation of an object, activity of one infant on behalf of another, and appreciation of the actions performed by one infant for another.

Findings

Positive emotional behavior — This is the earliest form of positive behavior but even so requires the prior development of emotional receptivity. All early positive relations seem to be the result of adult identification and older children and adults are the preferred objects. Positive emotional play occurs by chance when a child is near by or approaches and some common stimulus is understood, or when one child responds to the actions of another. About the age of 17 months one child may join another engaged in a recognized form of play. This play is imitative at first but gradually changes. Brief, sporadic play becomes longer and groups become larger. After 21 months the increasing complexity of social behavior is emphasized by the beginning of dramatized play. (However, in the case of one child positive and complex positive forms predominated throughout; she introduced sequential behavior and themes in her positive interchange.)

Common manipulation of an object — This play begins with the understanding that the given activity is shared with others; it is modifiable and adaptive, and imitative of adults. The author emphasizes that children refuse to participate in play if they do not completely understand all the elements involved. Play may begin when one child, watching and comprehending the activities of another, either imitates or takes part in them, or when two children are simultaneously engaged in similar activities. Play becomes more sustained when the child is about 17 months of age. At this time children play in larger groups and seem to have common goals. They often play together with building blocks in the second year but do not try to build together until much later.

Activity of one child on behalf of the other — The strongest stimulus is understanding another's activity. This depends upon the child's personality and training. The development of sympathetic behavior is described in these terms.

Appreciation of another child's behavior — Appreciation of another's behavior usually does not develop to any extent in the second year and is rarely seen before 20 months of age.

OTHER NEW FORMS OF BEHAVIOR IN SECOND YEAR

Beginning in the second year the following changes take place: looking for social contact and joining in play are noted from the beginning of the second year; coordination of activity with others, helping one another, sympathy, and grudges and offended feelings appear from 17 months; forceful removal of objects is replaced by exchanges and requests from 18 months; moral judgments are passed from the age of 20 months; quarrels appear from 20 months; and imitation of adults in play begins at the end of the second year.

238

SANFORD, R. N., *et al.* *Physique, Personality and Scholarship; a Cooperative Study of School Children* (Society for Research in Child Development, Monograph, vol. 8, no. 2). National Research Council, Washington, D. C., 1943

Setting: School and clinic

Subjects: 48 children; age 5 to 14 years

Time span: 3 years

Meth. of obs. and test.: Physical and physiological studies — records of illness and dietary habits; physical measurements; physiological measurements of pulse rate, basal metabolic rate, autonomic nervous system variables, urinary creatinine, urinary sex hormones, osseous development, and coordination and related variables. Psychological studies — rating scale based on H. A. Murray's concept of needs, press, mode, and actones; school reports; interviews with teachers; teachers' questionnaires; Rorschach Test; Thematic Apperception Test

TREATMENT OF DATA (by M. M. Adkins and R. N. Sanford)

Graphic method — A technique of syndrome analysis is employed in this monograph. It is presented here because of its relative importance for other longitudinal research. Also the study presents deviations for standard error of the technique described. This method is used to select high or low variables when each variable is plotted against age and then correlated with age. Thus two extreme groups, the highs and the lows, are obtained for each age group. In a sense, by this method age can be disregarded and the child can be considered high or low; the experimenters used the data in this way. Every correlation coefficient represents the relationship of two variables when age is partialled out in this manner. This allows comparison of children of different age groups and avoids the problem of a wide age range.

Method for estimating the correlation coefficient — The coefficients are largely estimates of the product-moment coefficients based on highs and lows for the two variables in question. The procedure is suggested in studies by Kelley and Flanagan.* Kelley pointed out the optimum demarcation line for obtaining high and low groups at the twenty-seventh and seventy-third percentile. Flanagan has prepared a chart for estimating the product-moment correlation coefficient from the proportion of "successes" in the extreme groups.

Correlation of variables — Thus in order to correlate they partialled out age and picked the highest and lowest twenty-seventh percentiles. Then they determined the correspondence of the highs and lows with other variables. The value of the correlation coefficient was then found in a table constructed on the basis of Flanagan's chart for estimated values of the product-moment correlation in a normal bivariate population.

Grouping of the data (the syndrome) — By intercorrelating variables, groups of interrelated variables were identified and termed syndromes. An attempt was made to relate the syndromes of physiologic, anthropometric, and personality types.

Method of constructing syndromes — Each variable of a syndrome had to correlate +.30 or more with each of the other variables in the syndrome. If this permitted duplication, the correlation was raised to +.40. For each syndrome the highs and lows were selected and compared for likeness in the different variables.

Method of selecting syndromes — The aim was to develop syndromes which differed from each other. Nearly 100 syndromes were described and named in terms of a factor which it was hoped would significantly identify that factor. Actually they were clusters of variables which were related statistically and were selected because of theoretical or descriptive congruence.

* T. L. Kelley, The Selection of Upper and Lower Groups or the Validation of Test Items, *Journal of Educational Psychology*, 30:17–24 (1939).

J. C. Flanagan, General Considerations in the Selection of Test Items and a Short Method of Estimating the Product-Moment Coefficient from Data at the Tails of Distribution, *Journal of Educational Psychology*, 30:674–680 (1939).

Theories behind syndromes — The two concepts behind the syndromes were that variables are different measures of the same thing and that, owing to a dynamic interrelationship, a change in one variable of the syndrome will affect the others.

PHYSIOLOGICAL SYNDROMES AND RELATED FINDINGS (by R. Bretney Miller)

Observations on autonomic nervous system — There was a moderate correlation among flushing, sweating, odor, acne, skin stroking intensity, and palpable thyroid. This was termed "the parasympathetic response syndrome" which is more common after puberty.

Sex hormone excretion (by Ira T. Nathanson, Lois Towne, and Joseph C. Aub) — A rather striking change toward masculinity was noted spontaneously by parents and teacher in two boys, both of whom had had obviously feminine habits. This change was coincident with a rise in the androgen-estrogen ratio from an estrogen excess to androgen preponderance. In general, these sex changes began 5 years before puberty but in the 1½ years just before puberty more rapid changes took place.

Intercorrelations of basal metabolism, creatinine excretion, urinary sex hormones, and osseous development — Two syndromes were described from these variables. The first was the high metabolism-advancing osseous development syndrome; the second was the advanced muscle, hormone excretion, and osseous development syndrome.

Coordination and related variables — The variables of the coordination syndrome were flexibility, balance, large and small muscle movement, grace, and lack of extraneous movements.

PERSONALITY SYNDROMES AND RELATED FINDINGS (by R. Nevitt Sanford)

Findings (based on observation by teachers and staff)

Ratings of child personality — The children tended to be consistent in their behavioral manifestations over the 3-year period as judged by teachers' questionnaires. The teachers' ratings by questionnaire were in terms of actones which were translated into need rating and then compared with the staff's ratings of needs.

It was found that the staff rated consistently higher on socially nonacceptable behavior than did the teachers with whom there was prolonged interaction.

Variables increasing with age — Those variables which seemed to increase with age were need-blame avoidance, need-abasement, need-order, need-endurance, need-counteraction, and perhaps need-defendance. These apparently reflected the process of socialization.

Variables decreasing with age — These were need-acquisition, need-aggression, need-cognizance, need-impulsion, need-blamescape, and need-succorance. These also represented the process of socialization (with the exception of need-nurturance) and seemed to be the converse of those variables increasing with age. The author points out the possibility that need-nurturance, giving of affection, and help to others are more instinctive than we think. Also the decrease in need-cognizance apparently is due to the child's decreasing curiosity or an unwillingness to display his ignorance.

Need-sex and need-exhibition — These two needs showed a U-shaped curve which is a substantial corroboration of the latency period theory of psychoanalytic thought.

Male and female differences — Though these differences were discussed, the findings were not statistically remarkable.

Findings (based on personality tests)

Occurrence of needs in fantasy — Need-aggression and need-acquisition were most frequently expressed in psychological tests and they tallied −.30 with a rank order on their manifest behavior. From this it is apparent that the needs with greatest societal restrictions have the greatest fantasy expression. Needs low in fantasy and high in overt behavior were need-understanding, need-order, need-counteraction, need-blame avoidance, need-deference, need-construction, and need-sentience. Those needs high in fantasy and overt behavior were need-nurturance, need-achievement, need-affiliation, need-dominance, and need-cognizance.

Variations of fantasy needs with age — Need-abasement, need-infavoidance, need-blamescape, need-blame avoidance, need-counteraction, need-deference, and need-acquisition all showed a marked increase from 8 to 14 years of age. The needs showing a decrease at this age were need-passivity, need-excitance, need-sentience, need-nutrience, need-sex, need-construction, need-affiliation, and need-play. From 10 to 12 years of age need-acquisition, need-achievement, need-recognition, need-dominance, and need-construction were at their highest, while need-affiliation, need-succorance, and need-harm avoidance were at their lowest point.

Summary of age variations — The above pattern of needs correlated with the syndrome of self-confidence, self-assertion, impersonality, and practical action. Children aged 8 to 10 years were concerned with sensations and not self-assertion; those aged 10 to 12 years emphasized self-forwarding and neglected social feelings. The oldest children (age 12 to 14 years) were concerned with guilt and inferiority, and were closest to the youngest children in their social feelings.

Sex differences in fantasy behavior — Girls had a more even distribution of needs and showed more need-sentience, need-seclusion, need-affiliation, need-rejection, need-nutrience, need-cognizance, and need-play. Boys displayed more need-retention, need-abasement, need-blamescape, need-autonomy, need-achievement, need-harm avoidance, and need-understanding.

Correlations between manifest behavior and covert needs — Strong antisocial needs in fantasy were accompanied by strong controlling needs; the latter, being socially acceptable, were openly displayed. On the other hand, there was also a correlation between fantasies lowering the self and assertive behavior. Covert needs and manifest needs may point in the same direction, the latter being a distorted form of the former (e.g., covert need-achievement correlated +.22 with manifest need-exhibition).

239

SCHACHTEL, A. H. The Rorschach Test with Young Children, *American Journal of Orthopsychiatry*, 14:1–9 (1944)

Setting: Nursery school

Subjects: 10 subjects, but only 1 is presented as representative

Time span: 5 years; age 3 through 7
Meth. of obs. and test.: Rorschach Test

The author describes the use of Rorschach tests for extremely young children, pointing out that the protocols are uncompromising, personally significant, and dynamic. Responses and interpretations are given for various ages. She is of the opinion that in one case the core of the subject's personality was revealed at age 3 — i.e., his concern with power and size as well as his overall harmony and productiveness.

240

SEARS, F., and WITMER, H. Some Possible Motives in the Sexual Delinquency of Children of Adequate Intelligence, *Smith College Studies in Social Work,* 2:1–45 (1931)
Setting: Clinic
Subjects: 50 children with sexual delinquency; I.Q.'s over 85; median age, 14 years
Time span: Not stated
Meth. of obs. and test.: Case histories

Findings

Personality differences in boys and girls — While the boys were quiet, shut-in, and rather effeminate, the girls were aggressive and sophisticated and also had many traits associated with the opposite sex.

Gregariousness as a factor in sexual activity — The children were divided into those with few friends and those with many. The sexual activities of girls with few friends were traceable to grossly inadequate homes and to a desire for affection. Boys with few friends seemed to be impelled by mental conflict or by need for male approval or to be responsive to the curiosity of aggressive girls. The gregarious children seemed rather well adjusted emotionally. Their homes revealed poor parental examples, inadequate training, and lack of affection. They seemed to seek out a "fast" crowd and defy the customs of the larger social group.

241

SEARS, R. R., WHITING, J. W. M., NOWLIS, V., and SEARS, P. S. Some Child-rearing Antecedents of Aggression and Dependency in Young Children, *Genetic Psychology Monographs*, 47:135–236 (1953)

Setting: Research institute

Subjects: 40 preschool children

Time span: Not stated

Meth. of obs. and test.: Interviews with mothers; teachers' ratings; standardized doll play situation; direct 15-minute observations

Findings

Correlation between oral frustration and dependency — There was a positive association between oral frustration and dependency with weaning an important related factor, especially among girls. This dependency was more manifest in relations with the teacher than in relations with other children.

Correlation between current frustrations and dependency — There was a positive association between current frustration and dependency for girls. However, this form of dependency was more manifest in relations with other children than in relations with the teacher.

Correlation between punitive maternal relationship and dependency — A punitive maternal relationship correlated positively with male dependency and negatively with female dependency.

Correlation between toilet training and aggression — A low correlation was found between severe toilet training and aggression in boys. In the presence of a punitive maternal relationship there was a drop in overt aggression but not in fantasy aggression.

Authors' interpretations

The kind and amount of frustration and punishment experienced by the child largely determined the properties of his dependency and aggressive drive. The radical sex differences in the developmental processes of these drives are probably due to the differential identifications with the mother. In the maternal treatment of boys and girls there are deep and pervasive differences after one year.

242

Sewall, M. Two Studies in Sibling Rivalry; I — Some Causes of Jealousy in Young Children, *Smith College Studies in Social Work,* 1:6–22 (1930)

Setting: Clinic and nursery schools

Subjects: 70 children; 40 boys, 30 girls; 24 children in nursery school and 46 who had been referred to the clinic because of behavior problems

Time span: Not stated

Meth. of obs. and test.: School and clinic records; home visits

Findings

Types of sibling rivalry — Of the 39 subjects who were definitely jealous of their younger siblings, 26 made physical attacks on their siblings, 9 showed definite personality changes at the time of the sibling's birth, 2 ignored them, and 2 denied having a sibling.

Significant factors in sibling rivalry — The first factor was the age of the child at the time of the younger sibling's birth, 18 months to 3 years being the danger period. The second concerned the number of children in the family, i.e., there was a small but steady decrease in the proportion of jealous children with an increase in family size. The third involved maladjustment in the family, of which by far the most important was inconsistent discipline. Four-fifths of the jealous children came from inconsistent homes, while only one-fifth came from consistent homes. Inconsistent discipline had a high correlation with other environmental conditions which seemed to contribute to jealousy such as poverty, overprotective mothers, negative fathers, and marital discord.

Insignificant factors in sibling rivalry — These factors were sex, intelligence, and whether or not the child was told that a sibling was expected.

243

Sherman, M. The Interpretation of Schizophrenic-like Behavior in Children, *Child Development,* 10:35–42 (1939)

Setting: Institution
Subjects: 17 children
Time span: 6 months to 4 years
Meth. of obs. and test.: Psychiatric examinations and treatments; physical and psychological examinations; Stanford-Binet Intelligence Scale

The most prominent symptom was emotional inadequacy in the children's adjustment to their environment. The subjects used poor judgment in their actual social contacts but had adequate judgment on tests. Their relations with other children were hindered by isolation, quarrelsomeness, and a reputation for being "odd." Although delusions were present, they were less fixed than in adult schizophrenics. Mannerisms and compulsive behavior were frequent.

244

SHIRLEY, M. Development of Immature Babies during Their First Two Years, *Child Development*, 9:347–360 (1938)
Setting: Research institute
Subjects: 63 children
Time span: 6 to 18 months
Meth. of obs. and test.: Pediatric and developmental examinations; anthropometric measurements

Findings

Length of retardation — Babies under 4 pounds were retarded at least 1 month up to the age of 18 months, but those weighing 4 to 5 pounds overtook the normal child by the age of 9 months.

Characteristics of premature babies — Premature infants were more retarded in manipulative development than in intellectual grasp and social responsiveness; 10% were accelerated and 10% feeble-minded. These babies had more nervous mannerisms. The environment of a premature infant tends to be unusual because of the association of prematurity with several other factors, e.g., primiparity, twinning, and advanced age of the mother.

245

Shirley, M. A Behavior Syndrome Characterizing Prematurely-born Children, *Child Development,* 10:115–128 (1939)

Setting: Research institute

Subjects: 3 groups of children — 65 children, age 6 to 30 months; 30 premature children, age 3⅓ to 6 years; 4 prematurely born children who had been intensely observed 10 years previously

Time span: Variable

Meth. of obs. and test.: Developmental examinations; medical histories

Findings

At the preschool age prematurely born children showed a behavioral syndrome characterized neurologically by keen auditory and visual sensitivity; lingual, manual, motor, locomotor, and sphincter control difficulties; hyperactivity or sluggishness; and in a few instances by tremor. Emotionally the subjects tended to have short span of attention and to be easily distracted, irascible, stubborn, shy, aesthetic, and overdependent on the mother. The syndrome was found more often in boys than in girls.

Author's interpretations

The author suggests that a comparison with normal controls might show whether these traits are specifically attributable to prematurity and to what extent other factors are involved.

246

Shirley, M., and Poyntz, L. Development and Cultural Patterning in Children's Protests, *Child Development,* 12:347–350 (1941)

Setting: Research institute

Subjects: 200 children; age 2 to 7 years

Time span: 7 years

Meth. of obs. and test.: Records of verbal protests during physical examinations

Findings

Chronological development — The call for "Mama" or "Daddy" comes at 2 years of age; this protest is nonadaptive since the parents were not there and it declines with age. Negative protests such as "No" and "I don't want to" appear earlier in girls, but specific protests such as "That hurts" and "All done?" appear about the same time in boys and girls. Belligerence is common to boys and girls at age 3. Whereas knowledge of the procedure leads to specific worries and anticipation in girls at age 3, it does not affect boys until age 3½. "I won't cry" is used by both at age 3½. Resignation and acceptance ("Are you going to?") and a modification ("Not too long," etc.) appear at age 5. At 7 years there are nonspecific responses such as "Ouch" or "Oh." Cultural influences are observed in the differences between sexes, e.g., in general girls tend to make an appeal while boys try to display bravery.

Summary of development — There is an unmistakable developmental trend in these children's protests from a nonadaptive appeal for parental aid at the beginning to self-reliant negativism, then to specific suggestions for getting out of a situation. From these the children advance to veiled excuses, threats, demands, and protests against specific procedures; even later they made resolute attempts to comply with the social code of not crying, indicated resignation, and finally resorted to an involuntary outcry. This progress is obviously toward the more socially approved expression of protest.

247

SHIRLEY, M., and POYNTZ, L. The Influence of Separation from the Mother on Children's Emotional Responses, *Journal of Psychology*, 12:251–282 (1941)

Setting: Research institute

Subjects: 199 children; age 2½ to 7½ years

Time span: 4 years

Meth. of obs. and test.: Diary-type records of each child's day at the research institute

The authors found that many children do suffer varying degrees of anxiety upon separation from the mother. Since this anxiety diminishes with age, the authors attribute it in part to physical immaturity. Separation anxiety is more frequent in boys; the authors suggest that this developmental immaturity in boys as compared with girls is the biological basis for the stronger mother-son tie. Anxiety arises largely from the child's fears that his needs will not be met; this may be allayed by a sympathetic, competent mother-surrogate. Prolonged fears which persist through the day and at repeated examinations are attributed to fears of rejection and loss of the mother's love. The authors state that the child's self-confidence and independence depend "on his having experienced warm and wise maternal care."

248

Shirley, M. M. *The First Two Years; a Study of Twenty-Five Babies*, vol. II (Institute of Child Welfare, Monograph Series, nos. 6–8). University of Minnesota, Minneapolis, 1933

Setting: Home and hospital

Subjects: 25 babies

Time span: 2 years

Meth. of obs. and test.: Mothers' records; developmental and pediatric examinations; observations in home; interviews with mothers; specific test situations; records of traits

This study attempted to cover all aspects of the first two years of childhood. In general the data were based on prolonged and detailed observations, reactions to test situations, the mothers' records, and other incidental items recorded by the observers. Valuable personality sketches of each child supplemented the objective data. Computations were made on the basis of traits and the individual score of each baby and trait was converted into a percentage of the median score. Changes in a trait were considered real only when they ran counter to the developmental curve of that trait. Individual personality was judged by the criteria of all-pervasiveness, pattern, permanency, and the possibility of development or change. The finding of traits that remain relatively constant during the first two years (when growth and change are the rule) was considered significant, and for this reason as well as others the infant was found to be a useful subject for the study of personality.

Findings

Differences at birth — The differences that remained consistently present from birth onward were considered to be inborn. Most notable in this respect was the tendency toward marked irritability or placidity.

Trends in development — All babies conformed to the developmental trend in all traits but individual differences were marked. They also tended to maintain traits which were harmonious and consistent with those of their families. In general babies who developed personality traits early performed thereafter at a higher-than-average proficiency level.

General patterns — One was a pattern of high general achievement with proficiency in speech and few nonadaptive responses; the other was exactly the opposite. The children differed in the rate of acquiring speech, their need for it, and their uses of it. This language development was in general compatible with other traits as they appeared in the examination and observations.

Author's interpretations

The author indicates that definite personalities do exist in babyhood when the incidental behavior of a child is traced and considered quantitatively. It corroborates the pattern of personality indicated by reactions to the psychological examination.

249

SHIRLEY, M. M. Children's Adjustments to a Strange Situation, *Journal of Abnormal and Social Psychology,* 37:201–217 (1942)

Setting: Research institute

Subjects: 181 children

Time span: 2 years

Meth. of obs. and test.: Ratings of each child's adjustment to the research center

These children were part of a complex longitudinal study and various other data are referred to.

Findings

Proportion of adjustment — One-half of the children maintained a consistent level of adjustment; of these two-thirds had consistently good adjustment. In the inconsistent group regression was the most common form of inconsistency.

Factors influencing adjustment — One factor significantly related to poor adjustment in the examining room was an unwholesome maternal attitude. Illness had little effect apart from the family's reaction to it. The group who began visits at 2 to 2½ months of age adjusted better than the group who began later, but this was not highly significant statistically. There was no correlation between adjustment and sibling number or position; some deterioration of adjustment was evident when the family constellation was changed by the birth of a sibling.

250

SIEGEL, M. G. The Diagnostic and Prognostic Validity of the Rorschach Test in a Child Guidance Clinic, *American Journal of Orthopsychiatry,* 18:119–133 (1948)

Setting: Clinic

Subjects: 26 children

Time span: 1 year

Meth. of obs. and test.: Rorschach Test; clinical histories

Findings

Comparison of tests — On the first tests there was a discrepancy in 10 cases between clinical and Rorschach impressions; at the second test disagreement persisted in only 3 cases. According to the psychiatric diagnosis, all 3 cases in which there was disagreement were primary behavior disorders; but according to the Rorschach diagnosis, 1 was a psychosis and 2 were psychoneuroses.

Rorschach tests — The tests were useful in confirming or contradicting psychiatric diagnosis, revealing psychotic trends which became clinically manifest in the treatment process, and gauging the effectiveness of therapy. Clinically, favorable response and accessibility to therapy seemed to depend on the presence of 8 Rorschach traits which correlate with certain facets of the personality structure: refusals (an awareness of personal disturbances which may result in a temporary reorganization of the personality in the direction of flight), F.C. (affective adaptability), W% (ability to solve problems and attain recognition), Fc (a capacity for tact and emotional discernment), H (interest in others), T.L. (readiness to accept new ideas), O (capability for original thinking), and F+% (an intact ego). The 4 signs associated with an unfavorable response to therapy were O-, CF-, Shading Shock, and CF.

Author's interpretations

The author feels that these signs should be considered valid only for a group with approximately the same characteristics as the group under study.

251

SILVERMAN, B. The Behavior of Children from Broken Homes, *American Journal of Orthopsychiatry,* 5:11–18 (1935)

Setting: Agency

Subjects: 138 children from broken homes

Time span: Not stated

Meth. of obs. and test.: Social work case histories; physical and psychiatric examinations; psychometrics; two weeks of general observations

Findings

Only 25% of the subjects were overtly abnormal or antisocial in their behavior. Factors present in broken homes included physical illness or death (36%), mental illness (23%), economic pressure (13%), sex delinquency (63%), other types of delinquency (62%), and serious neglect or cruelty (36%). No significant correlation was found between homes broken by delinquency and incompatibility of the parents and the behavior of the children.

Author's interpretations

When children from such homes develop behavior problems,

they are probably attributable to the subtler emotional relations within the family group rather than to identification with the overt delinquencies of the parents.

252

Simon, A. J. Rejection in the Etiology and Treatment of the Institutionalized Delinquent, *Nervous Child*, 3:119–126 (1944)

Setting: Institution

Subjects: 110 boys and 39 girls

Time span: 1 year

Meth. of obs. and test.: Social work case histories; rating scales of response to treatment

Findings

Characteristics of delinquent children — A high incidence of parental rejection was found among the delinquent children in this study. Such children were infantile or distorted emotionally — e.g., 17% of the boys and 35.5% of the girls had psychoneurotic character structures. Narcissism and lack of object-feeling relationships were evident in 80% of the boys and 64% of the girls.

Post-treatment adjustment — After 1 year of treatment 18% of the boys were adjusting normally in their behavior, 34% showed improvement in all 5 indices of personality, and 11% showed improvement in none.

Author's interpretations

The author discusses the problems arising from an institution's attempt to provide interpersonal relationships, and the findings reaffirm the extreme problems proceeding from the therapy of the psychopathic personality.

253

Skeels, H. M., Updegraff, R., Wellman, B. L., and Williams, H. M. *A Study of Environmental Stimulation; an Orphanage Preschool Project* (Studies in Child Welfare, vol. 15, no. 4). University of Iowa, Iowa City, 1938

Setting: Institution

Subjects: 2 groups of orphanage children, 59 experimentals and 53 controls; factors equal in each group except for attendance at preschool

Time span: 3 years

Meth. of obs. and test.: Binet Intelligence Scale; Merrill-Palmer Scale; Little-Williams Language Achievement Scale; Smith-Williams Vocabulary Test; General Information Test; Vineland Social Maturity Scale; timed observations; Berne Rating Scale for Social Behavior; McCaskill Motor Achievement Test

Findings

Intelligence — From results of intelligence tests and retests, it was evident that preschool attendance made a "decidedly effective contribution toward better intellectual development." It seemed apparent, too, that children of average ability in the control group could fall back into the feeble-minded category according to both I.Q. ratings. Although the preschool group had more general information than the controls, both were below the average for their age. (Based on scores for two nonorphanage nursery schools.)

Language — Language development was retarded in both groups. It was believed that contributory factors included a lack of association with adults and deficiency of experience on which to build verbal expression, both of which were characteristic of the orphanage. The preschool experience was only slightly effective in modifying language achievement.

Social competence — After a large initial gain by the preschool group, both groups slowly lost in the social quotient; however, a year's development consistently separated the two groups. That the preschool groups did not continue to gain was, the authors thought, partially the fault of the school which could not fulfill all the children's needs.

Behavior — In the 24 preschool children observed throughout the year, laughing and smiling, watching, and talking to another child decreased significantly whereas making advances to another child, accepting advances, and continuing activities longer increased. In general, aimless withdrawal and anxious behavior decreased while purposive, controlled, social behavior increased. Berne social ratings on 38 of the experimental group (5 times, 1934–1936) indicated a therapeutic trend for extremes of social behavior to move toward the middle range.

Motor performance — The preschool group progressed faster than the control group as a result of better environment and opportunity.

254

SKODAK, M., and SKEELS, H. M. A Follow-up Study of Children in Adoptive Homes, *Pedagogical Seminary and Journal of Genetic Psychology,* 66:21–58 (1945)

Setting: Iowa Child Welfare Research Station

Subjects: 139 children placed in adoptive homes before 6 months of age; mean placement age, 3 months

Time span: 7 years

Meth. of obs. and test.: Kuhlmann-Anderson Intelligence Test; Stanford-Binet Intelligence Scale

Findings

I.Q. test and results — The average I.Q. results were 116 at 2 years, 112 at 4 years, and 113 at 7 years. Between the second and third tests changes of 20 points or more were still found but with less frequency than between the first and second.

Influence of foster homes — The subjects' development corresponded to expectations for natural children in homes which were similar economically and socially to those of foster parents. They continued to be markedly superior to their true parents in intelligence and there was no indication that they would revert to a lower level. Children in relatively superior foster homes remained somewhat superior to other foster children but differences decreased at the age corresponding to entry into school. Children with mentally defective mothers could not be differentiated from the total group on the basis of intelligence.

255

SLATER, E. *Studies from the Center for Research in Child Health and Development, School of Public Health, Harvard University; II — Types, Levels, and Irregularities of Response to a Nursery School Situation of Forty Children Observed with Special Reference to the Home Environment* (Society for Research in Child Development, Monograph, vol. 4, no. 2). National Research Council, Washington, D. C., 1939

Setting: Nursery school and home

Subjects: 18 boys; 22 girls

Time span: 2½ years

Meth. of obs. and test.: Records of routine, directed, and undirected activities using timed observations; records of emotional disturbances; narrative account of each child's first day in the nursery school

RESPONSE OF YOUNG CHILDREN TO THE NEW ENVIRONMENT OF AN OBSERVATION NURSERY SCHOOL

Case histories are presented to illustrate types of response and methods of adaptation.

Findings

Activity ratings — Marked contrasts were observed in the children's tempo of activity during the first few days of school. These differences were evident in the number of activity changes, number of toys investigated, and the number of contacts with adults and other children. Many changed greatly in tempo during the 6-week period. There were 8 children who demonstrated and 4 who tended toward an "accelerated" type — i.e., one whose score was high on response at first and then decreased sharply. There were 7 children who demonstrated and 6 who tended toward an "inhibited" type — i.e., one whose score was low on response at first and then increased considerably. Of 20 children who returned for later observation, 17 still showed the same tempo of activity. Of the original 40, 15 had no pattern — i.e., they fluctuated constantly in nature, number, and tempo of activities. These were called "alternating." Of the latter group, 14 returned for follow-ups and still showed "alternating" tendencies;

9 ultimately adopted a more uniform pattern, 5 moved toward the "accelerated" group, and 4 toward the "inhibited."

Behavior ratings — During the first few days tics, tears, postural tensions, facial anxiety, dreamy watching, and group rejection were frequent; at subsequent visits these wore off. Crying was least in the inhibited group; dreaminess, group rejection, tics, and tensions were least in the accelerated group.

Characteristics of reaction types — There were no age or sex differences in the distribution of the three reaction types nor in general adjustment. However, none of the highly intelligent children were accelerated, 8 were inhibited, and 5 were alternating. The author suggests that high intelligence leads to increased perception of strangeness and thus increased the difficulty of adjustment.

Environmental factors — A somewhat higher proportion of alternating children had oversolicitous mothers and came from homes with unfavorable health factors. More crying and tics were present in children from moderate homes. There was no relation between adjustment types and the presence of siblings or additional adults at home.

INTELLECTUAL RESPONSES

Data were presented which placed each child in the appropriate stage of mastery at successive observation periods.

Findings

Language development; recognition of form differences; color matching; picture responses; block building; paints, crayons, and clay; rhythmic responses; and the use of scissors were the eight items in which intellectual factors seemed prominent.

MOTOR AND OTHER PHYSICAL RESPONSES

Various stages of mastery were observed and charted. Studies for gross motor control comprised mastery of stairs, Kiddy-kars, and slides. Fine skills were observed in relation to a mastery of buttons and pouring from a pitcher.

SOCIAL AND EMOTIONAL RESPONSES

Findings

Types of contact — The 19 children aged 2½ made 54 friendly contacts (median number) in 125 minutes, 30 children aged 3 made 41, 30 children aged 3½ made 118, and 22 children aged 4 made 144. The large increase at 3½ years was attributed chiefly to speech. Thus the percentage of associative play in 125 minutes was 9% at 2½ years of age, 14% at 3 years, 20% at 3½ years, and 30% at 4 years. Maternal contacts increased from 19 at 2½ years of age to 26 at 4 years; physical contacts increased from 2 at 2½ years to 10 at 4 years of age.

Speech — The 17 children aged 2½ years talked to other children 44% of the time, 28 children aged 3 years talked 50% of the time, 25 children aged 3½ years talked 56% of the time, and 21 children aged 4 years talked 66% of the time. When speech to adults was excessive, it suggested the need for attention and reassurance; no age trends were seen. Excessive speech to self suggested self-preoccupation or introversion.

General observations — Passive watching was a personal rather than a social response and age seemed to be less important than familiarity with nursery experience. Group participation in this setting seemed to have no relation to age. Emotional disturbances (violent manifestations of emotion) tended to disappear at higher age levels.

IRREGULARITIES IN THE LEVELS OF RESPONSES WITH REFERENCE TO CHANGES IN HOME ENVIRONMENT

Sample case histories are presented as well as an appendix illustrating sample records.

Findings

Irregularities in response by the same child — Two-thirds of the group was at least moderately consistent and one-third decidedly inconsistent in responses to separate intellectual items at a given age. After consistency had been defined as 6 out of 8 responses in the same quartile at successive visits, it was found that two-thirds were consistent in responses to intellectual items at successive ages; one-third was irregular. With respect to the large and fine motor items, as in the case of intellectual items, about one-third of the group showed marked inconsistencies and irregularities both at given and at successive ages.

Consistency of various responses — Intellectual responses in the nursery school and in the psychological testing room indicated considerable conformity; however, some children showed striking disparity and all these had "decidedly poor emotional poise." There was less consistency between items and at the various visits with respect to social and emotional responses than with respect to intellectual and motor responses. However, there was a relative consistency with respect to such items as tics, acceptance of group activities, and negativism.

Symmetry and asymmetry in the various responses — The averages of each child for the various items were computed; those in the same or adjoining quartiles were called regular. Wide disparity indicated asymmetry of development or response; 13 out of 33 children showed this degree of asymmetry. Of this group 8 showed little or no change between visits (asymmetry was characteristic) and the other 5 fluctuated considerably. Possible factors associated with these 13 irregular children were changes in the home environment — e.g., a new baby, illness, unemployment, separation of parents, etc. Of the 17 children who experienced changes in their homes while in the nursery school, 15 were irregular and 10 of these were markedly so; of the 13 most irregular only 2 had not passed through recent home alteration. Chronic home situations did not seem to be associated with such irregularities. Nor was there any difference between boys and girls. There was a tendency for children who were irregular in the nursery school to be irregular in physical growth.

256

Smalley, R. E. Two Studies in Sibling Rivalry; II — The Influence of Differences in Age, Sex, and Intelligence in Determining Attitudes of Siblings toward Each Other, *Smith College Studies in Social Work*, 1:23–40 (1930)

Setting: Clinic

Subjects: 27 pairs of children; each pair were siblings from a 2-child family and both children were studied by the clinic

Time span: Not stated
Meth. of obs. and test.: Clinic case histories

The children were divided into 3 groups: 11 pairs were markedly jealous, 9 pairs had attitudes of protection-dependency, and 7 pairs seemed mutually friendly.

Findings

Jealousy — A larger proportion of jealousy was present in the 5 pairs of sisters (60%) than in the 9 pairs of brothers (44%) or in the 13 pairs of mixed siblings (30%). Jealousy predominated when there was a great difference in intelligence between the siblings, especially if the older was the duller child. If the duller child had a special talent, this helped alleviate the situation. Siblings of the same sex were more apt to develop jealousy than those of opposite sexes though this was a less significant factor than the differences in intelligence.

Sibling relations — Protective attitudes toward a sibling were shown by 38% of the mixed pairs, 33% of the brothers, and 20% of the sisters. More than any other factor, a wide age difference led to a protective attitude on the part of the elder child. Similarity of age and intelligence enhanced friendly relations between siblings.

Parental preferences — Parental preferences were not usually affected by marked differences in intelligence or ordinal position but were more likely to occur when siblings were of the same sex. In 17 cases there was a slight tendency for parents to prefer the duller child. Mothers were more apt to prefer girls, whereas fathers preferred boys.

257

Sontag, L. W. Effect of Fetal Activity on the Nutritional Status of the Infant at Birth, *American Journal of Diseases of Children,* 60:621–630 (1940)

Setting: Fels Research Institute
Subjects: Not stated
Time span: Not stated

Meth. of obs. and test.: Polygraph records made by the mothers during pregnancy; birth weights

Findings

Differences in fetal activity — When infants were compared, there were differences of 1,000% in fetal activity. These differences were not consistently related to size — some light babies were active and some inactive; some inactive babies were light and some heavy; but no heavy baby was active.

Some causes of light infants — Maternal waste products (a result of fatigue) sometimes increased fetal activity which in turn produced light infants. Two case histories showed that maternal emotional upsets were also associated with fetal activity and comparatively light infants.

258

Sontag, L. W. The Significance of Fetal Environmental Differences, *American Journal of Obstetrics and Gynecology,* 42:996–1003 (1941)

Setting: Fels Research Institute

Subjects: Not stated

Time span: Pregnancy and postpregnancy

Meth. of obs. and test.: Polygraph records made by the mothers during pregnancy

Findings

Fetuses that were active during the last two months of pregnancy were small at birth and showed more advanced motor development during the first postnatal year. Prolonged emotional disturbances of the mother sometimes produced autonomic dysfunction resulting in early feeding disturbances.

259

Sontag, L. W., and Richards, T. W. *Studies in Fetal Behavior; I — Fetal Heart Rate as a Behavioral Indicator* (Society for Research in Child Development, Monograph, vol. 3, no. 4).

National Research Council, Washington, D. C., 1938
Setting: Fels Research Institute
Subjects: 30 mothers; 29 unmarried and 1 married
Time span: Last 6 months of pregnancy
Meth. of obs. and test.: Heart rates calculated by stethoscope

Findings

The fetal heart rate decelerated and its variability increased as the fetus approached birth; after birth its rate was further reduced and more variable. As the fetus matured, maternal activity had a greater effect on the heart rate. Vibration (the most potent cardio-accelerator) and bodily activity in utero accelerated the heart rate while maternal smoking could accelerate or decelerate the rate; fetuses sensitive to vibration could be insensitive to smoking and vice versa.

260

Spitz, R. A. Hospitalism; an Inquiry into the Genesis of Psychiatric Conditions in Early Childhood, in: *Psychoanalytic Study of the Child* (vol. I). International Universities Press, New York, 1945
Setting: Private homes, nurseries, and foundling homes
Subjects: 164 children; 34 in private homes and 130 in 2 institutions in different countries
Time span: 1 year
Meth. of obs. and test.: Psychiatric anamneses; Hetzer-Wolf Test; special experiments; motion pictures and written protocols

Findings

Developmental quotient — The children in the first three environments were generally well developed and normal at the end of the first year. The fourth group started fairly high on the quotient but deteriorated. The "restriction of psychic capacity" was seen as a progressive process. By the end of the second year, the foundling home children's developmental quotient had decreased to 45, corresponding to a mental age of about 10 months. The nursery group curve did not deviate significantly from the normal although it dropped at two points, i.e., at 6–7 and 10–12 months.

Background and environment	*Developmental quotient (average)*	
	First 4 months	*Last 4 months*
Parents' home		
Professional class	133	131
Village population	107	108
Nursery	101.5	105
Foundling home	124	72

Foundling home children — These children showed mental and physical signs of "hospitalism." Of the 88 children under 2½ years of age, 23 died. Of the remaining children, 26 were between 1½ and 2½ years of age. All of them were incontinent, only a few could eat alone, and only 2 could speak a few words and walk alone. The foundling home children had better hereditary equipment than the nursery children, but they had no mothers and only an "eighth of a nurse." As soon as they were weaned (at 4 months), their few human contacts diminished and their development fell below normal.

Nursery children — These, in contrast to the foundling home children, were enterprising, sociable, and normally acceptable children. The nursery mothers were delinquent girls who were experiencing severe narcissistic trauma because their pleasure sources were limited. The child became representative of their sexuality, an object of pride, and a phallic substitute; mothers were in constant competition as to who had the "better baby."

Author's interpretations

Other deprivations of the institutionalized child are minor in comparison with his isolation from perceptual human stimulation, especially since mother-child relations compensate for other deprivations.

261

Spitz, R. A., and Wolf, K. M. Anaclitic Depression; an Inquiry into the Genesis of Psychiatric Conditions in Early Childhood, in: *Psychoanalytic Study of the Child* (vol. II). International Universities Press, New York, 1946

Setting: Institution
Subjects: 123 infants
Time span: 12 to 18 months
Meth. of obs. and test.: Hetzer-Wolf Test

Findings

A syndrome was found in 19 children who had good relations with their mothers for the first 6 months of life and then lost contact with her for a period of at least 3 months. The infant began to withdraw and cry a great deal. There was a drop in the developmental quotient at 9 months and a loss of responsive contact with the examiner. This syndrome is characterized by withdrawal and lack of responsive contact; apprehension, sadness, and weeping; dejected expression; slow movement and stupor; insomnia; waning appetite and weight loss; retarded reaction to stimuli and retarded development.

262

Spitz, R. A., with Wolf, K. M. The Smiling Response; a Contribution to the Ontogenesis of Social Relations, *Genetic Psychology Monographs*, 34:57–125 (1946)

Setting and subjects:

Race	*Nursery*	*Private home*	*Foundling home*	*Delivery clinic*	*Village*	*Total*
White	57	15	21	12	..	105
Colored	39	..	..	..	..	39
Indian	..	..	48	33	26	107
Total	96	15	69	45	26	251

Time span: 54 children, birth to age 20 days; 12 children, age 20 to 60 days; 132 children, age 20 days to 6 months; 13 children, age 3 to 6 months; 39 children, age 6 months to 1 year; 108 children, age 20 days to 1 year

Meth. of obs. and test.: Separate tests were made by male and female examiners to define the onset of smiling. During early testing they looked at the child full in the face, turned to a profile position, and then back if the child stopped smiling.

Findings

Results — The results indicated a general onset of smiling at 2 months; indiscriminate smiling diminished at 6 months. There were no differences in social or racial classes.

Response	Number of days			
	0–20	*21–60*	*61–180*	*180–360*
Smile	..	3	142	5
No smile	54	141	3	142
Total	54	144	145	147

Conditions which provoke smiles — One set of conditions found to be successful in provoking a smile was the presence of a configuration and the absence of emotion. For example, an exaggerated grimace proved sufficient stimulus in all but one case. It was also found that when the examiner turned to a profile or covered one eye, the infant stopped smiling. Thus the child's smile had nothing to do with the recognition of an emotion in the adult's face. Another set of conditions which provoked a smile was the presence of a configuration and the absence of human attributes. For example, a mask with movement elicited response from all but two children and a scarecrow was also successful. But a large selection of other stimuli were inadequate, e.g., the child did not smile at a large series of familiar and unfamiliar toys.

Clinical impressions — Early smiling was concomitant with advanced development, both general and intellectual; early distinction between mask and face occurred in precocious children. Infants who didn't smile between 3 and 6 months of age either lacked perceptive development (prognostic sign of deficiency) or had some psychiatric involvement in the emotional-social phase of development. Persistence of smiling at 6 months occurred in institutionalized children with developmental retardation.

Authors' interpretations

The authors conclude that a full-face position, display of both eyes, and motion are essential to eliciting a smile. They infer "that the presence of the smile in the second trimester of life is a necessary (though not sufficient) indicator of emotional homeostasis. . . ." They also feel that the infant's smile is an indication of his response to his relationship with his mother. This response is due in human beings to the fact that the infant can eye the mother while nursing.

263

Spitz, R. A., and Wolf, K. M. Autoeroticism, in: *Psychoanalytic Study of the Child* (vols. III and IV). International Universities Press, New York, 1949

Setting: Institution

Subjects: 196 institutionalized children, 26 of whom were eliminated because of age factor

Time span: 1 year (first year of life)

Meth. of obs. and test.: 200 hours of observation per child; interviews with nurses and mothers; Rorschach Test for 30% of the mothers

Autoerotic behavior was examined in terms of four categories: rocking, genital play, fecal play, and no activity.

Findings

Distribution of autoerotic categories — Of 170 children, 87 rocked, 21 played with the genitals, 16 played with feces, and 66 engaged in no such activity. Rocking was most prominent at 6 to 8 months of age and genital play at 10 to 12 months. A control group of 17 privately reared children showed that 16 manifested genital play in the first year and this began 2 months earlier than the institutionalized group. Another group of 61 foundlings reared without emotional relations had only 1 child who exhibited genital play.

Maternal patterns associated with autoeroticism — Mothers having children who rocked were extraverted, alloplastic, infantile, and inconsistent in child care; they also had poor control of aggression and could not bind tension. In the group of mothers having children with problems of fecal play, there was a significant percentage of psychotics with depression. Characteristic of this group was a change from prolonged oversolicitude to rejection and deep-seated ambivalence sometimes expressed in actual physical attacks on the child. Those mothers having children with genital play had close relations with their children and maintained consistent attitudes without extremes of libidinal need or aggression.

264

STEVENSON, S. S. Paranatal Factors Affecting Adjustment in Childhood, *Pediatrics,* 2:154–161 (1948)

Setting: Not stated

Subjects: 226 children in the Harvard public health longitudinal study

Time span: Not stated

Meth. of obs. and test.: Rating scales of adjustment based on longitudinal study; rating scales of physical behavior during the first 2 days of life

Findings

Of the infants who had bad paranatal conditions, 39% were later rated as maladjusted. Of infants who had good paranatal conditions, 19.2% were later rated as maladjusted. This was a highly significant difference but specific factors could not be isolated.

265

STEWART, A. Excessive Crying in Infants; a Family Disease, in: *Problems of Infancy and Childhood* (Transactions of the Sixth Conference, 1952). Josiah Macy, Jr., Foundation Publications, New York, 1953

Setting: University of Washington Research Institute

Subjects: 21 babies and their parents; 12 with excessive crying

Time span: Not stated

Meth. of obs. and test.: Psychological tests and psychiatric evaluations of parents; observations of color, secretion, and swelling of mucosa of infants; eosinophil counts; barium swallow and G.I. series; records of pulse respiration; developmental studies; infants' reactions to being separated from mothers, being isolated in a crib, and being held and played with; mothers' diaries; prenatal interviews with mothers

A description of case material is presented. Crying is attributed apparently to familial tension. The concept of crying as a response is briefly considered and is evaluated as a complex response varying in significance with reference to time of onset, accompaniment of other behavior, etc. Certain psychosomatic insights are indicated by the correlation between crying and nasal discharge. Excessive crying results in a mutually dissatisfying relationship for mother and child. The parents are "perfect parents" wanting their babies to be perfect.

266

STEWART, A., *et al.* Excessive Infant Crying (Colic) in Relation to Parent Behavior, *American Journal of Psychiatry,* 110:687–694 (1954)

Setting: Research institute

Subjects: 18 infants from 13 families; 10 boys, 8 girls

Time span: Prenatal contact with mothers; 6 months' study of the infants from birth

Meth. of obs. and test.: Psychological tests and psychiatric evaluations of parents; observations of color, secretion, and swelling of mucosa of infants; eosinophil counts; barium swallow and G. I. series; records of pulse respiration; developmental studies; infants' reactions to being separated from mothers, being isolated in a crib, and being held and played with; mothers' diaries; prenatal interviews with mothers

Findings

Complete Group

Definition and description — Crying was described as a response to internal tension resulting from unsatisfied needs or inappropriate external stimulation. The parents' behavior was perceived by the infant through sensory and proprioceptive systems. Infants who cried excessively did not develop as much security in interpersonal relationships as those who cried less. Among criers there were also many other symptoms.

Frequency and duration of crying — The frequency, intensity, and duration of crying varied widely in the first 2 weeks of life. Though 4 babies cried less than 1 hour per day, 6 cried from 4 to 11 hours per day. After 2 weeks, the infants could be divided into 3 groups.

Group I (8 infants)

Chronological development — During the first 3 months of life these infants cried excessively or episodically (for 90 minutes or more) over a period of at least 2 weeks. This crying was not related to physical discomfort. At 1 month of age crying began in the late afternoon and was associated with increased muscle tension, sweating, and nasal discharge. The spells lasted from 2 to 7 hours. At the age of 2 months crying was more intense and was accompanied by frenzied activity; the children had anxious expressions and symptoms of excessive gas, which was confirmed by x-ray. These infants also showed an increased growth rate during the crying period and 4 infants showed a lag after cessation of crying. This first group also had frequent upper respiratory infections, rashes, and falls.

Maternal factor — The mothers of this group were all anxious, i.e., overactive toward the baby during the first month and inadequate after the first month. All mothers had conflicts about acceptance of the feminine or maternal role, dependency needs, and rivalry with the child or husband. These conflicts were expressed in their neglect or overcompensating attention toward the child. All breast-fed their children, but only 2 continued more than 3 months. They offered the children solids before 3 weeks of age and handled the baby excessively. The fathers of this group were passive and nonsupporting.

Group II (4 intermediate infants)

These infants responded variably to their parents. Crying increased with neglect or stimulation as in Group I. The mothers showed less hostility toward the child and more dependent rivalry than those in Group I; fathers were like those in Group I.

Group III (6 infants)

Infants cried less from 3 weeks of age; when present, crying was related to obvious stimuli. They had little muscle tension, no G.I. trouble, and normal growth. The mothers had little anxiety and their behavior was appropriate to the infants' apparent needs. There were fewer conflicts about parenthood.

267

STUTSMAN, R. Irene; a Study of the Personality Defects of an Attractive Superior Child of Preschool Age, *Pedagogical Seminary,* 34:591–614 (1927)

Setting: Nursery school

Subject: Preschool girl, Irene, who attended the Merrill-Palmer School from the age of 3 years 2 months to 5 years 1 month

Time span: 2 years

Meth. of obs. and test.: Case history; interviews with parents; observations in school; physical examinations; intelligence, performance, and psychological tests; follow-up visits at school

This is an example of the case history records of the Merrill-Palmer School.

Findings

Child's background — Irene was an attractive preschooler of high intellect and good health who came from a comfortable, above-average home and had a younger brother. Her parents were intelligent and agreeable. The father's chief defect was his tendency to evade issues and procrastinate; the mother's chief defects were a sense of personal infallibility and severity in judging others.

Child's personality — The child was self-conscious, demanded a great deal of attention, and was very adroit in getting it. She showed extreme evasiveness to commands, requests, and tasks; yet she was bright, observant, imaginative, had an excellent memory, and profited from experience. The personnel in the school persistently worked to curb her showing off but were thwarted by the home's fostering of this tendency; however, some improvement seemed evident. Follow-ups showed she continued to avoid realities she disliked.

Author's interpretations

Further study will determine the influence of these undesirable traits acquired from her parents upon the child's later life.

268

STUTSMAN, R. Constancy in Personality Trends, *Psychological Bulletin,* 32:701–702 (1935)

Setting: School

Subjects: 140 children

Time span: 3 years

Meth. of obs. and test.: Rating scales

Findings

Scores based on ratings have been made into profiles which in many cases continue to be consistently uniform in pattern. The patterns of these profiles suggest the possibility of classification into personality types. These include a repressed, relatively quiet type with excessive emotional control and an active, clever, versatile type with little emotional control. Also there is the stable, well-adjusted, capable person; the weak, colorless person; and the unstable, fluctuating person.

269

SUARES, N. D. Personality Development in Adolescence, *Rorschach Research Exchange,* 3:2–12 (1938)

Setting: High school

Subjects: 77 boys and 21 girls; 4 groups — 21 retested boys, 21 retested girls, 21 high school boys, 14 clinic boys

Time span: Variable

Meth. of obs. and test.: Rorschach Test

Findings

Retested boys — There was a definite increase in movement responses (introversiveness). Of 5 introversive boys, 3 indicated more M and 2 showed no change; of 7 extratensive boys who remained extratensive, 3 did so despite a decrease in color and 4 showed an increase in M and ΣC. The total number of responses varied greatly in the retest, as did the D's and Dd's.

Retested girls — Less change was seen in the personality structure of the girls. There was no reversal of M:ΣC as was the case with the boys, and girls tended to become extratensive which was the reverse of the boys' trend. Between the two tests, the total number of responses varied with a general tendency to increase; W's, D's and Dd's varied markedly but also tended to increase.

Comparison of retested boys and girls — Changes in Erlebnistyp were seen in the group of boys aged 15 to 18 years where 6/7 showed an introversive trend with increased M and decreased ΣC. The younger group showed only a decrease in ΣC indicating that the increase in M was manifest only after age 14. Girls aged 12 to 14 years remained constant in M and increased in ΣC and the number of extratensive subjects increased from 5 to 11. The median number of W's for girls was 5 and for boys was 6; for the D's, the difference was in the opposite direction. While boys tended to be abstract, girls were more apt to be practical. Though both groups decreased in A% and increased in H%, the boys changed more than the girls. However, differences of sex tended to equalize.

Comparison of all groups — Girls gave more responses, D, Dd, and S; boys gave better associations. Girls indicated a richer, more varied affective life due to their higher scores in the affective sphere, but they were more superficial and extraverted than boys. With reference to M, high school boys had an average of 1.80, clinic boys 1.66, and random boys 1.43. F% varied little in all groups. High school boys had a more developed human interest and were less stereotyped (low A%, high H%) than clinic boys; the latter also gave the fewest W responses. All groups were nearly equal in P, though the clinic boys were lowest; this can be attributed to poorer adaptation to surroundings.

270

SWAN, C. Individual Differences in the Facial Expressive Behavior of Preschool Children; a Study by the Time-Sampling Method, *Genetic Psychology Monographs,* 20:557–650 (1938)

Setting: Nursery school

Subjects: 15 boys, 10 girls; age 27 to 49 months

Time span: 6 months

Meth. of obs. and test.: Time sampling with Becker Time Marker

Findings

The Pyknic types tended to be more sociable, extraverted, and expressive. Boys had more solitary vocalization but an equal number of social vocalizations. Children used more variant forms of expression at the end of the day. There were correlations between I.Q. and facial motility. Hyperactivity was positively related to facial motility and negatively related to hand-to-face activity.

Author's interpretations

The various types of nonsocial vocalization seem to be related and occur in the same child. The author notes that there were marked and characteristic differences between the different types of children and amount of facial motility.

271

TEAGARDEN, F. M. Change of Environment and the I.Q., *Journal of Applied Psychology,* 11:289–296 (1927)

Setting: Institution

Subjects: 2 orphans with I.Q.'s of 73 and 77, respectively

Time span: 6 years

Meth. of obs. and test.: Family, school, and state records; repeated I.Q. tests; Mooseheart staff observations and appraisals

These children came from a markedly defective social and economic situation and it was hoped that their I.Q.'s, 73 and 77 respectively, would be raised by the improvement of their social milieu.

Findings

The I.Q.'s remained unusually stable despite the change of

environment, the average deviation being 1.68 for one and 1.92 for the other. There was, however, a considerable improvement in social, physical, and domestic habits and skills.

272

TERMAN, L. M., BURKS, B., and JENSEN, D. *Genetic Studies of Genius; Vol. III—The Promise of Youth; Follow-up Studies of a Thousand Gifted Children.* Stanford University, Stanford, Calif., 1930

Setting: Not stated

Subjects: 1,000 gifted children

Time span: 6 years

Meth. of obs. and test.: Teachers' ratings; school records; Stanford-Binet Intelligence Scale; interviews with parents and children; artistic specimens; Terman Group Test; Thorndike Test; questionnaires

Only Volume III of this three-volume set is recorded in this bibliography since most of the longitudinal findings are contained in it. The book is divided into four parts.

PART I

This part presents a statistical study of the follow-up data. Most of the children selected were accelerated in their age-grade status; their I.Q.'s were 140 or above.

Findings

The children had been successful in a wide variety of extracurricular activities as well as their school work. The boys were found to be as masculine as boys with average I.Q.'s; however, the girls in this group were significantly less feminine than the average. Of the group, 95% wanted to go to college, but due to economic or social reasons only 80% could plan to go. The higher the I.Q., the more difficult the child's social adjustment in later life. However, in general, the group as a whole had no greater social problems than a comparative average group.

Authors' interpretations

The data based on this group suggests that without the support of their high intelligence their talents tended to fade.

PART II

This section presents a series of case studies, many of which are of unusual interest.

PART III

This contains a study of the literary productions of some of this group.

PART IV

This part summarizes the study and includes an outline of suggested future research, general considerations of the group, and a composite portrait of the gifted child which is presented by Terman.

Findings

In general, the group comprised the top 4 of 1,000 children with respect to intelligence; there were twice as many boys as girls when selected by this criterion for the sample group.

Authors' interpretations

The composite gifted child comes from good genetic stock which is decreasing in fecundity, has siblings with an I.Q. average of 125, is healthier than children of lesser I.Q., and is not significantly more maladjusted than the average child. Boys tend to maintain their I.Q. whereas girls seem to drop somewhat. However, in general there is evidence of the continued permanence of superiority in these gifted children.

273

THEIS, S. V. S. *How Foster Children Turn Out* (Publ. 165). State Charities Aid Association, New York, 1924

Setting: Not stated

Subjects: 910 foster children

Time span: Variable through age 18 years

Meth. of obs. and test.: Social evaluations

The author used the criterion of a well-adjusted child as one "capable of managing himself and his affairs with ordinary prudence."

Findings

Of the 797 subjects whose condition could be ascertained, 77% had made successful adjustment to society. The proportion of success was higher among the children placed in better homes (81%) and still higher among those placed before age 5 (86%).

274

THOMAS, D. S., LOOMIS, A. M., and ARRINGTON, R. E. *Observational Studies of Social Behavior*. Yale University, New Haven, 1933

Setting: Schools

Subjects: Not stated

Time span: 1 year and variable

Meth. of obs. and test.: One observer checked material and physical activity; another recorded time-sampling data of talking, physical contact, and coyness; motion pictures

The main attempt of this book is to describe techniques for the observation and analysis of social behavior in units. The units do not take into account motive or the overall situation. In connection with this study, however, repeated observations were made which indicated a rather low conformity among the relative positions of the same children in the group over a one-year period.

275

TRAMER, M. Tagebuch über ein geisteskrankes Kind, *Zeitschrift für Kinderpsychiatrie*

Vol. I Aug. 1934, pp. 91–97
I Oct. 1934, pp. 123–126
I Dec. 1934, pp. 154–161
I Feb. 1935, pp. 187–194
II Apr. 1935, pp. 17–28
II Aug. 1935, pp. 86–90
II Oct. 1935, pp. 115-124

[See translation by H. Bruch, and F. Cottington, A Detailed

Diary of the Early Years of a Schizophrenic Child, *Nervous Child,* 1:232–249 (1942)]

Setting: Home

Subject: 1 boy

Time span: 12 years

Meth. of obs. and test.: Diary by mother; medical observations

The diary consists of a detailed description of the psychomotor development of the child. There is no information about the mother-child relationship or about marital adjustment. The subject was a first-born boy who weighed 5½ pounds at birth.

Findings

Development up to 1 year of age — At 10 days the infant would suck in response to a smacking sound and at 2 weeks he was already turning toward the breast. It was noted early that the child was distracted by the clock striking and that he had found his fingers — by 25 days he was turning his head toward sound, was able to fixate his gaze, and seemed to prefer lying on his back. By 35 days his mother had noticed him laugh. At 7 weeks he seemed to recognize his mother. By 8 weeks he had begun to suck his fingers so constantly that they were sore. It was clearly noticeable at the end of 17 weeks that his sleep was cyclical rather than periodical and that often he slept quite poorly. Although he had previously developed the ability to lift his head, he suddenly lost this at 20 weeks of age. From this point on there seemed to be a retardation of motor development. The child paid very careful attention to his shadow by the time he was 10 months old. He failed to initiate activity and seemed to be trying to throw his foot out of the crib. By 11 months his motor reactions were so disturbed that he had to be lifted like a doll.

Development from 1 to 2 years — By age 1 year he seemed quite moody and easily overexcitable and he had temper tantrums and engaged in a great deal of whirling play. When he began to walk he carried a tape measure which he thought supported him. By the age of 18 months he had developed a specific negativistic reaction which he displayed by refusing to say "Please." He liked books and did not play well with ordinary toys. By 19 months of age he seemed clearly unable to play by himself in any of the usual ways. He would cry by the hour, especially at night, and perseverate on final syllables; he was markedly timid in attempting any motor action. Almost all his play was whirling or talking to his shadow by the time he was 23 months. He liked books, had a passion for leafing through them, and always wanted to carry one with him. He seemed unable to get up or sit down alone.

Development from 2 to 3 years — By 2 years 2 months he would hold on to his mother when walking. He talked mostly to himself, constantly repeated and mixed words, and confused basic concepts such as "up and down." His negativism was such that he would rather not eat than say "Please" to his mother. At the age of 2 years 5 months the child's language was interlarded with neologisms. He had become quite hostile to strange or "dirty" children. At the age of 2 years 9 months he was quite mute with strangers and all his play was turning, twisting, and whirling. The child was obviously withdrawn, isolated, and grossly disturbed by the age of 2 years 10 months.

Author's interpretations

The author summarizes these early symptoms of childhood schizophrenia in such psychiatric terms as verbigeration, stereotypy, autism, phonographism, perseveration, blocking, and perverse affectivity.

276

Troup, E. A Comparative Study by Means of the Rorschach Method of Personality Development in Twenty Pairs of Identical Twins, *Genetic Psychology Monographs,* 20:461–556 (1938)

Setting: School

Subjects: 20 pair of twins

Time span: 6 months

Meth. of obs. and test.: Rorschach Test

The author considered personality development as divided into three phases — tempo, quality, and direction of development.

Findings

The twins in the study tended to vary in at least one (and often more) of the three aspects mentioned above. There was no high degree of resemblance between twins of a pair but all the twins in the group had some basically similar personality trends. Only one pair of twins showed a similar trend of development.

Author's interpretations

The author points out that the data indicate "how widely different two people with basically similar personality make-up can become in the course of development." The intertwin contingency coefficient of .40 suggests a substantial environmental component. The I.Q. coefficient was .90.

277

TRYON, C. M. *Evaluations of Adolescent Personality by Adolescents* (Society for Research in Child Development, Monograph, vol. 4. no. 4). National Research Council, Washington, D.C., 1939

Setting: School clubs (part of the California adolescent growth study)

Subjects: Approximately 125 subjects studied longitudinally; age 12 to 15 years

Time span: 3 years

Meth. of obs. and test.: Ratings on 20 personality traits gathered from the "Guess Who" test

Findings

Comparison of boys and girls — Between the ages of 12 and 15 years, girls' values changed vitally whereas boys' changed only slightly and represented merely a shift in emphasis.

Characteristic behavior of girls — At the age of 12 years, girls associated quiet, sedate, nonaggressive qualities with friendship, likeableness, good humor, and attractiveness. Behavior approved by adults was admired; tomboyishness was tolerated. At age 15 admiration for the subdued, lady-like prototype ceased and many criteria for the ideal boy became highly acceptable for girls, e.g., extraverted activity and good sportsmanship. Other new and important ideals were "being fascinating to the opposite sex" and the ability to organize and maintain activities involving both sexes.

Characteristic behavior of boys — At the 12-year level the boys' ideal was a skillful, daring, personable boy who had leadership traits and was without feminine characteristics such as tidiness or marked conformity in the classroom. At age 15 skill, aggressiveness, and fearlessness were still traits of prestige but defiance of adult standards was associated with immaturity. Personal acceptability was now emphasized and indicated some rising heterosexual interests.

Author's interpretations

The author's observations indicate that many girls around age 13 exhibit behavior "suggestive of emotional upheaval" which probably coincides with changes in the system of values described above. This behavior is characterized by desultory interests in the objective environment, excessive affection and narcissism, and disorganized activity. Boys show no similar period of disorganized activity.

278

Veo, L. A Personality Study of Six Adolescents Who Later Became Psychotic, *Smith College Studies in Social Work,* 1:317–363 (1931)

Setting: Hospital

Subjects: 6 patients

Time span: 3 to 10 years

Meth. of obs. and test.: Records of child psychiatric clinic and later medical records

The six case histories are presented in summary form.

Findings

In most of these cases the consulting authorities made no early, clear prediction of the child's future psychosis. However, a serious maladjustment was recognized in 5 cases and it was stated that without some correction of the environmental situation, grave difficulties would arise. In the 3 cases of schizophrenia which developed the adolescent personality showed a constellation of those traits usually associated with the diagnosis — e.g., social withdrawal, introspection, and marked preoccupation with personal difficulties. In the 3 cases of psychosis with psychopathic personality, the characteristic traits of instability and irresponsibility appeared at an early age. In 3 cases treatment was solely environmental and only partially carried out, but in the other 3 cases treatment was both psychiatric and environmental and still unsuccessful.

Author's interpretations

It appears that early prediction of psychosis is difficult and mental breakdowns cannot always be attributed to a lack of social-psychiatric treatment during adolescence.

279

WALCOTT, E. Daydreamers; a Study of Their Adjustment in Adolescence, *Smith College Studies in Social Work,* 2:283–335 (1932)

Setting: Clinic

Subjects: 17 withdrawn children who had been treated in various ways

Time span: 2 to 8 years

Meth. of obs. and test.: Interviews and correspondence with families; social work case histories; data from an earlier study

Detailed case histories are included in this study. The boys were similar physically (i.e., underweight, small, awkward, and poorly coordinated) and all had been referred to the clinic before age 12. On the other hand, the girls were not similar and all but 2 were referred after age 14. Follow-up studies reveal the data described below.

Findings

Two boys (aged 7 and 13 when originally referred) were socially well-adjusted and had normal interests and contacts. Another pair of boys (aged 11 and 17 when originally referred) had changed greatly but still seemed somewhat more self-conscious than the average. A boy aged 11 and 2 girls aged 15 and 22 when originally referred showed less withdrawal from reality at the follow-up but limited their friendships to persons much like themselves in personality and interest. Three girls (aged 10, 14, and 15 at the time of original referral), who were considered to be of dull-normal intelligence, had given up their excessive daydreaming but had found simple work and had little contact with other people at the time of follow-up. Little change was found in 4 boys; 3 girls were apparently worse.

Author's interpretations

The author suggests that preadolescence with its accent on physical prowess is a period more difficult for boys than for girls who are more likely to retire into daydreams at adolescence when sexual attractiveness is valued. She points out that age and intelligence were not related to later adjustment, except in the intellecually subnormal group. The children who failed to give up their seclusive behavior were those who lived under the greatest and most persistent home disadvantages such as sibling rivalry, parental rejection, and overprotection. The author thinks that the prognosis for resolving the emotional difficulties of a seclusive, daydreaming child by means of socio-psychiatric treatment is favorable only when the child has some source of affection in his family.

280

WALLACE, R. A Study of the Relationship between Emotional Tone of the Home and Adjustment Status in Cases Referred to a Traveling Child Guidance Clinic, *Journal of Juvenile Research,* 19:205–220 (1935)

Setting: Clinic

Subjects: 248 cases for which "home tone" and adjustment were judged; 161 cases for which only adjustment was judged

Time span: 4 years

Meth. of obs. and test.: Case histories; interviews with psychiatrist, psychiatric social worker, and psychologists; medical and school reports; gross ratings of adjustment and "home tone"

Findings

Of the 248 children whose homes were judged, 63% were from harmonious and 37% from unharmonious homes. Of the two groups of children, 81% from harmonious and 71% from unharmonious homes showed improvement. Thus the children from harmonious homes tended to improve more than those from unharmonious homes. Of 308 follow-ups, 26% showed excellent adjustment, 46% good adjustment, 7% fair adjustment; 13% were unchanged; 3% were worse; and 5 were in institutions for defectives or delinquents.

281

WALSH, M. E. The Relation of Nursery School Training to the Development of Certain Personality Traits, *Child Development,* 2:72–73 (1931)

Setting: Research institute and nursery school

Subjects: 22 nursery school children; 21 non-nursery school children

Time span: 6 months

Meth. of obs. and test.: Taussig Industrial Classification of Parents; Bonham-Sargent Scale

Findings

The nursery school children became less inhibited, more spontaneous, and more socialized with training. They showed increased initiative, independence, self-assertion, self-reliance, and a greater increase of curiosity and interest in their environment.

Author's interpretations

The author concludes that all of the generally accepted, socially desirable traits increased in the nursery school group, while the less desirable traits increased in the control group. The former were more orderly and their health habits were more defined and more numerous than the control group. The author infers that the superiority of personality traits in the nursery school children as shown by statistical validity of the scores on the Bonham-Sargent Scale is due to the influence of the social force of a large group of children who had constantly to adjust to one another.

282

WARD, A. The Only Child; a Study of 100 Cases Referred to a Child Guidance Clinic, *Smith College Studies in Social Work*, 1:41–65 (1930)

Setting: Clinic

Subjects: 100 only children referred to guidance clinic; 73 boys and 27 girls; mean age, 8.2 years

Time span: Variable

Meth. of obs. and test.: Social work case histories; data from teachers

At the time of referral only children were notably younger than the clinic children as a whole (8.2 and 11.2 years respectively) and ranked higher in intelligence (109.8 and 103.3 respectively). The behavior problems of only children were quite similar to those of the clinic children, except that less stealing, lying, and truancy were found in the former group. When compared with a control group of families with three children, only children showed more restlessness, overactivity, crying, nail-biting, school difficulties, and unpopularity.

Author's interpretations

The author believes that the overconcern of the parents, the greater intelligence of parents, and the possibility that only children adjust as they get older and therefore do not require referral may account for the lower age average of only children. The latter's greater intelligence may be accounted for by their superior homes, older parents, and more adult association. The smaller incidence of stealing, lying, and truancy among only children is attributed to their age at the time of referral, their more sheltered environment where desires are oversupplied, and their limitation of contacts. The author feels that certain problems are unavoidably associated with being an only child — i.e., oversatisfaction of the need for affection; lack of competition for parental attention; too much parental ambition; insufficient contact with other children; and excessive contact with adults, leading to self-consciousness, negativism, and other pathological defenses.

283

Washburn, R. W. A Study of the Smiling and Laughing of Infants in the First Year of Life, *Genetic Psychology Monographs,* 6:397–537 (1929)

Setting: Research institute

Subjects: 15 children; 9 girls, 6 boys

Time span: From 8 to 52 weeks of age

Meth. of obs. and test.: Experimental situation with observations on 15 major items; pictures and charts

The author uses case histories to illustrate the three personality groups described below and includes a summary which differentiates the various types of behavior associated with smiling.

Findings

Differences between laughing and smiling — Laughing occurred later than smiling and when first seen was more stereotyped in form; laughter could not be differentiated at various age levels. "Coy" smiling was seen in two girls and was the only detectable sex difference. More individual differences were noted in the frequency of response than in the particular form of response. The frequency of smiling and laughing did not correlate with chronological age, mental development, or physical condition. Nor was a relationship noted between the physical type and the type of expressive behavior detected.

Stimuli of laughing and smiling — The methods of stimulation which elicited smiling also elicited laughter in the early weeks. The stimulus required physical proximity and often abrupt physical contact. Smiling without stimulation was almost never seen although this has been frequently reported by other authors. The methods for eliciting laughter which were best for most ages were in general best for most of the children. By the age of 52 weeks smiling was conditioned to certain stimuli which could be described to a certain extent, but laughter seemed to remain an unconditioned response.

Personality groups — By analyzing overt expressive behavior three personality groups were determined. The first included the ambi-expressive children who seemed to smile and laugh as much as they cried; the second covered the risor-expressive who tended to smile and laugh more than the average; and the third included the depressor-expressive who seemed to cry more than they smiled or laughed.

284

WASHBURN, R. W., and PUTNAM, M. C. A Study of Child Care in the First Two Years of Life, *Journal of Pediatrics,* 2:517–536 (1933)

Setting: Well-baby conference

Subjects: 59 babies

Time span: 2 years

Meth. of obs. and test.: Pediatric and developmental examinations; interviews with mothers to obtain data on sleeping, feeding, etc.

Findings

Child care — In general the mothers were using approved methods of child care, but they seemed to be unaware of the wide individual differences to be expected among infants. They were eager for information and receptive to suggestions, but they rarely sought help and incorporated advice into their program of child care only when their own experience emphasized its value.

Results of conference — Of the 46 children studied sequentially, 27 presented practically no problems; 13 had successive transient, minor problems; and 6 had persistent difficulties. Prognosis seemed good for 33, questionable for 9, poor for 2. A preliminary period of difficult adjustment in the first 2 or 3 months of life was noted. During this time the child adjusted to the rhythmic conditions of living and the mother to her new responsibilities and emotional state. The psychologist's role became defined in regard to the well-baby conference.

285

WENGER, M. A. The Measurement of Individual Differences in Autonomic Balance, *Psychosomatic Medicine,* 3:427–434 (1941)

Setting: Research institute
Subjects: 62 children
Time span: Not stated
Meth of obs. and test.: 7 measurements of autonomic function

During the period of a year 48 children proved to have consistent scores. Basing his analysis on these scores, the author offers a scale of seven measurements as a tentative method for measuring individual differences in autonomic balance and formulates a multiple regression equation derived from factor analysis.

286

WENGER, M. A. A Further Note on the Measurement of Autonomic Balance, *Psychosomatic Medicine,* 5:148–151 (1943)
Setting: Research institute
Subjects: 81 children; age 6 to 13 years
Time span: 3 years
Meth. of obs. and test.: 7 measurements of autonomic function

A significant autonomic factor is revealed by the combined data of the factor analysis. An adequate new regression equation labeled the "normative regression equation for the estimation of autonomic balance" is described.

287

WENGER, M. A., and ELLINGTON, M. The Measurement of Autonomic Balance in Children; Method and Normative Data, *Psychosomatic Medicine,* 5:241–253 (1943)
Setting: Research institute
Subjects: 40 boys, 40 girls
Time span: 3 years
Meth. of obs. and test.: 7 measurements of autonomic function

The higher the scores on these seven tests, the more the parasympathetic predominated over the sympathetic. Normative scores are given and charts are presented; raw scores can be converted by means of the latter. The author describes a new regression equation called the "normative regression equation for the estimation of autonomic balance." He considers this to be more reliable than three previously employed regression equations based on rounded scores.

288

WHITE, M. A., and WILLIAMS, H. M. The Approach-Withdrawal Pattern in the Social Behavior of Young Children, *Pedagogical Seminary and Journal of Genetic Psychology,* 54:73–84 (1939)

Setting: School

Subjects: 53 kindergarten children; age 55 to 68 months

Time span: 1 year

Meth. of obs. and test.: Records of contacts and degree of reaction

Findings

The mean for the group was 51.9 contacts in the fall and 62.6 in the spring. This difference suggested a growth in social behavior. Exclusiveness for the group was fairly low in that the children formed no strong cliques. There were substantial changes during the year in the number of contacts made by each child and a slight tendency for the extremes to become more balanced. The distinct changes in responses toward preferred individuals during the year was proven by the fact that identity with the four most frequently contacted children was only 27.7%. The authors point out that their findings were highly variable.

289

WILE, I., and DAVIS, R. The Relation of Birth to Behavior, *American Journal of Orthopsychiatry,* 11:320–334 (1941)

Setting: Clinic

Subjects: 500 children with behavior problems; age 13 to 15 years

Time span: Not stated

Meth. of obs. and test.: Data on birth procedures and family; I.Q.'s; categories of behavior traits

Findings

Position in the family — While first-born children formed the largest group seen in child-behavior clinics, no position in the family was free from special problems of adjustment.

Method of delivery — A far larger proportion of undesirable behavior traits (aggressiveness, submissiveness, peculation, sibling conflicts, infantile family relationships, school difficulties, tics, and physical ills) was found among the spontaneously delivered children (Group A). Far more hyperactivity (100%) and more superior I.Q.'s were found in the instrument-delivered children (Group B). Yet, the latter group seemed to show a general reduction in "personality energy."

Authors' interpretations

Since undesirable behavior traits were not associated with nonspontaneous delivery, the authors conclude that "shocks of birth" do not create a persisting anxiety which lead either to neurosis or a general form of sensitivity. However, they suggest that physical trauma may be indicated by the correlation of instrument delivery with hyperactivity and decreased "personality energy."

290

WILE, I., *et al.* The Continuity of the Neurotic Processes, *American Journal of Orthopsychiatry,* 4:49–72 (1934)

Setting: Clinic

Subjects: 50 children who presented problems at the clinic before 10 years of age (Group A); 50 children who presented problems at the clinic between the ages of 11 and 16 (Group B)

Time span: Group A seen from median age 8½ to 15; Group B seen from median age 13 to 15

Meth. of obs. and test.: Specific techniques not stated

Findings

In Group A, 23 of the 50 subjects were free of neurotic symptoms in adolescence, 11 were partially free, and 16 retained their symptoms. In Group B, 10 subjects did not have histories of neurotic symptoms, 15 had histories of at least 1 symptom, and 24 had histories of continuous neurotic behavior. In all, 40 of the 100 children manifested a definite continuation of neurotic behavior, while 27 showed some evidence of disconnected but similar behavior, and 33 showed no neurotic continuity at all.

Authors' interpretations

The authors conclude that neurotic predispositions manifest themselves under varying conditions at different age levels. Latent neurotic symptoms may suddenly appear at adolescence if conditions are present that are conducive to disturbed behavior. On the other hand, the sudden disappearance of neurotic behavior is believed to be merely a temporary discontinuation rather than a permanent cessation. In the present series continuity and shifts were both observed. The authors believe that maturation plays an indefinite part in these shifts; they feel that early psychiatric treatment has the advantage of breaking the continuity of the neurotic process. They point out that high intelligence is associated with continuity and that self-inhibiting (flight-type) behavior is more likely than aggressive (fight-type) behavior to continue through the preadolescent period into adolescence. Disturbed behavior which is manifested by motor habits is less likely to continue. In general, however, prognoses of the continuity of neurotic processes are impossible.

291

Witmer, H., Hough, E., Foley, P., Lewenberg, M., Figge, M., Brunk, C., and Freeman, M. Studies in Maternal Over-Protection, *Smith College Studies in Social Work,* 2:181–282 (1932)

Setting: New York Institute for Child Guidance

Subjects: See individual articles

Time span: Variable

Meth. of obs. and test.: Social work case histories

This is a series of seven papers designed to test Dr. D. M. Levy's theory that there is a distinct syndrome called "maternal overprotection" caused by certain specific conditions in the mother's life. These conditions include a long period of anticipation and frustration during which a desire for a child is thwarted, factors in the child which threaten his survival (e.g., defects, illness), sexual incompatibility with the husband, social isolation, emotional impoverishment in early life, development of dominating characteristics through the assumption of undue responsibility in childhood and marriage, and thwarted ambitions. The overprotecting attitude, according to Dr. Levy, is manifested by excessive contact with the child (e.g., sleeping together, excessive fondling, and keeping the child in sight), prolongation of infantile care, prevention of child's development of independent behavior, and lack or excess of maternal control (e.g., overindulgence or overdemand). Two groups were compared in this series, one composed of mothers who were overprotective and the other of mothers whose behavior showed little evidence of overprotection or rejection. The study also considered the rejecting mother who was defined as one whose behavior toward her child indicates a strong dislike. All the parents were patients of the New York Institute for Child Guidance. Statistical data, interpretations, and case abstracts were included in the original papers which are briefly summarized here.

INTRODUCTION (by H. Witmer)

SOME FACTORS IN THE ETIOLOGY OF MATERNAL OVER-PROTECTION (by E. Hough)

A group of 32 overprotective mothers was compared with an equal number of controls who were neither overprotective nor rejecting in regard to traits believed to be associated with the overprotective attitude. The children were similar with respect to age, intelligence, maternal nationality, and religion.

Findings

The overprotected children were more frequently boys and only children. The overprotective mothers exceeded the controls in the number of unhappy childhoods, in the quantity of early responsibilities, and in the degree of hazard to health presented by having children; but the controls exceeded in the number of mothers having a dominant role in marriage. Those having thwarted ambitions and social isolation were equally common in both groups.

EARLY RESPONSIBILITY AND AFFECT-HUNGER AS SELECTIVE CRITERIA IN MATERNAL OVER-PROTECTION (by P. Foley)

One hundred mothers were chosen at random to determine the relationship, if any, between their own childhood experiences and their attitudes toward their children.

Findings

An overprotective mother characteristically experienced in her own childhood a lack of affection coupled with responsibility; a rejecting mother received affection but lacked any responsibilities; and a neutral mother had both affection and responsibility. Regarding marital adjustment, both overprotective and rejecting mothers were more likely to be dominant in marriage than neutral mothers but dominance did not seem to be related to a particular type of childhood experience. However, an overprotective attitude on the mother's part was often accompanied by, or associated with, unsatisfactory marital adjustment.

MARITAL DISHARMONY AS A FACTOR IN THE ETIOLOGY OF MATERNAL OVER-PROTECTION (by M. Lewenberg)

Two groups of 45 mothers (overprotective and control) were compared for 14 factors thought to be influential in marital discord.

Findings

There was a consistently greater positive incidence of family disharmony in the overprotective group, particularly in social and sexual maladjustment, disagreement over discipline and the desire for children, interference from relatives, and economic dissatisfaction. All these Dr. Levy called "causal" factors in marital dissatisfaction and overprotection. Factors which Dr. Levy termed "incidental" had a similar incidence in the two groups. They were thwarted ambition, domination by mother, and the fact that love was not the motivating factor in marrying. Combinations of the positive factors were much more frequent in the overprotective group.

SOME FACTORS IN THE ETIOLOGY OF MATERNAL REJECTION (by M. Figge)

A group of 35 rejecting mothers was compared with an equal number of controls. The children did not differ in age, family size, ordinal position, I.Q., or physical factors. There were a few problems present which were thought to be results rather than causes of rejection.

Findings

The mothers who rejected their children had unhappier childhoods, greater social frustrations (through marriage), more marital incompatibility, thwarted ambitions, slightly inferior economic status, and more sex and pregnancy fears. As opposed to 6 nonrejecting mothers, 33 rejecting mothers did not want children. The rejecting mothers were shocked by and ill during pregnancy; the controls were not. The biggest single factor characterizing the rejecting mother was the lack of emotional ties to parents in her own childhood.

THE EFFECTS OF MATERNAL OVER-PROTECTION ON THE EARLY DEVELOPMENT AND HABITS OF CHILDREN (by C. Brunk)

Case histories were presented and a comparison made of 30 overprotective and 200 nonoverprotected patients. The overprotective mothers were somewhat older; overprotected children were somewhat brighter.

Findings

There were no statistically significant differences between the groups in the age of learning to walk, talk, and establish bladder control, but the overprotected child tended to learn earlier. The latter could not be attributed to superior intelligence. There were no marked differences between the two groups in eating, sleeping, and illnesses. Significant differences were found in the length of breast-feeding — 40% of the overprotected children and only 7% of the controls were breast-fed 12 months or more. Economic status and nativity were excluded as factors.

FACTORS ASSOCIATED WITH LENGTH OF BREAST FEEDING (by M. Freeman)

A study of 526 case records showed that 11% had been breast-fed less than 1 month and 32% for 12 months or more. A group of 100 mothers was taken from both extremes of the scale and compared; the choice of women compared favorably with other cross-sectional studies regarding the selection of subjects.

Findings

Neither the sex of the child nor the intelligence of the mother was associated with the length of the nursing period. The significant factors included ethnic background, the number of children in the family, and the mother's attitude. There was a slight tendency for foreign-born mothers to nurse their children 12 months or more. Mothers having an only child tended to nurse 1 month or less, while those having 3 to 4 children nursed them 12 months or more. One-half of those nursing 1 month or less were rejecting mothers; two-thirds of those nursing 12 months or more were overprotective.

292

WOLBERG, L. R. The Character Structure of the Rejected Child, *Nervous Child,* 3:74–88 (1944)

Setting: Hospital

Subjects: 33 children; 24 hospitalized

Time span: Variable

Meth. of obs. and test.: Psychotherapy with children; interviews with mothers

Findings

Background — Most of the mothers were immature and unstable and in some cases they were actually psychotic.

Types of rejection — Two main types of parental rejection were differentiated: rejection and hostility (involving antagonism and criticism), and neglect.

Reactions to rejection — The most common reactions were detachment, compulsive dependency, power striving, aggression, emotional immaturity, and inordinate preoccupation with sex. Internal conflicts arose when several of these reactions developed simultaneously. In such cases aggression usually predominated. Generally, the children showed less self-esteem. In some cases rejection had a stimulating effect on self-development in that the child seeking security attempted to be independent and master his environment.

Author's interpretations

The author believes that parental attitudes are far more important than the economic status of the family or the physical condition of the home. He concludes that rejection is most catastrophic during the first year of life when the child is helpless, tensions are high, and the need for motor and sensory stimulation is great. The worst effects are from inconsistency and contradictory attitudes because the child becomes confused and is unable to develop a sense of values. The mother-child relationship is most important in personality development; if the child is rejected, he usually makes defensive, reparative efforts to supply the lacking gratification. The rejected child may never develop a repressive system capable of inhibiting immediate pleasure promptings.

293

Wolf, K. M. Observation of Individual Tendencies in the First Year of Life, in: *Problems of Infancy and Childhood* (Transactions of the Sixth Conference, 1952). Josiah Macy, Jr., Foundation Publications, New York, 1953

Setting: Hospital, home, and research institute

Subjects: 2 infants

Time span: 4–5 months prenatally through first year of life

Meth. of obs. and test.: Interviews with mothers; well- and sick-baby care; Rorschach Test and Wechsler-Bellevue Intelligence Scale given to mothers; postnatal period with detailed records; Gesell and Hetzer-Wolf tests; detailed observations of mother-child interaction

The two cases of Christina and Steve are presented at length. These cases illustrate how the child's behavior develops out of the parent-child relationship and reflects the conscious and unconscious needs of the parent. They also illustrate the "equipment" of the child at birth and the day-to-day handling by the mother.

294

WOLMAN, B. The Social Development of Israel Youth, *Jewish Social Studies,* 11:283–306, 343–372 (1949)

Setting: Not stated

Subjects: 4,000 boys and girls

Time span: 4 years

Meth. of obs. and test.: Interviews; personal diaries; questionnaires; general observations

The article includes a survey of the following aspects of social development at the prepubertal stage: the phenomena of classroom unrest, the formation of gangs, youth movements, expression and duration of friendship, solitude, and moral development. The primary reason for solitude at the pubertal stage is occasioned by difficulty in finding suitable friends and groups. This selectivity is an indication of social maturation. In the postpubertal period subjective attachment to friends and group opinions are replaced by objective values and social ideas and forms.

295

WOOLLEY, H. T. A Dominant Personality in the Making, *Pedagogical Seminary,* 32:569–598 (1925)

Setting: Merrill-Palmer School

Subject: 1 girl; age 3 years

Time span: 2 years

Meth. of obs. and test: Case history; interviews with parents; observations in school; physical examinations; intelligence and performance tests

This case is an example of the kind of longitudinal study done at the Merrill-Palmer School. Agnes's ancestors on both sides were adventurous, independent people with great initiative. Her mother attributed the child's domineering behavior to the fact that she was humored throughout infancy because of precarious health. At the age of 7 months she acted as though the universe revolved around her, and her parents' attitude was completely subservient. At school Agnes was seen as an executive in the making; she showed real ability and a flair for managing and planning.

While recognizing that these traits were assets that should be preserved, the school authorities tried to teach her, before it was too late, that other people have fundamental rights. It was felt that some improvement resulted from the school experience.

296

WOOLLEY, H. T. II — Case of Peter; the Beginnings of the Juvenile Court Problems, *Pedagogical Seminary and Journal of Genetic Psychology,* 26:9–29 (1938)

Setting: Merrill-Palmer School

Subject: 1 boy; age 4 years

Time span: 4 years

Meth. of obs. and test.: Case history; interviews with parents; observations in school; physical examinations; intelligence and performance tests

This longitudinal study by the Merrill-Palmer School illustrates the way in which delinquent tendencies get an early start. The author believes that Peter's delinquent behavior originated in his relations to his parents, who are described as emotional, vacillating, and ineffectual. They were preoccupied with their own affairs and lacked consistency and a high regard for truth in their handling of children. During five months at the school the child seemed to improve but after he left his intelligence score fell below average and he reverted to his career of truancy and stealing. The school experience was too brief to correct his habits of feeling, thought, and action. In the author's opinion Peter's unfavorable development could have been averted.

297

YARROW, L. T. The Role of Oral Deprivation and Gratification in Non-nutritive Sucking, *American Psychologist,* 6:305 (1951)

Setting: Child Research Council, University of Colorado

Subjects: 66 children; 28 boys, 38 girls

Time span: Not stated
Meth. of obs. and test.: Interviews with mothers

Findings

Children who had inadequate sucking opportunities in early infancy were the most severe thumb-suckers and persisted in this habit the longest time. The age of bottle-weaning was not related to the severity or duration of thumb-sucking. Children weaned late were more severe thumb-suckers and also showed more frustration at the time of weaning than children weaned early. Girls showed more severe reactions than boys.

INDEX

All numbers in this index refer to article numbers rather than page numbers.